DIALECTICS OF NATURE

? 1872 - 1882

Dialectics of Nature

BY FREDERICK ENGELS

TRANSLATED AND EDITED BY CLEMENS DUTT

WITH A PREFACE AND NOTES BY

J. B. S. HALDANE, F. R. S.

INTERNATIONAL PUBLISHERS

NEW YORK

COPYRIGHT, 1940, BY
INTERNATIONAL PUBLISHERS CO., INC.
PRINTED IN THE U.S.A.

Seventh Printing, 1973

SNB: (cloth) 7173–0049–0; (paperback) 7178–0048–2

CONTENTS

PREFACE

MARXISM has a two-fold bearing on science. In the first place Marxists study science among other human activities. They show how the scientific activities of any society depend on its changing needs, and so in the long run on its productive methods, and how science changes the productive methods, and therefore the whole society. This analysis is needed for any scientific approach to history, and even non-Marxists are now accepting parts of it. But secondly Marx and Engels were not content to analyse the changes in society. In dialectics they saw the science of the general laws of change, not only in society and in human thought, but in the external world which is mirrored by human thought. That is to say it can be applied to problems of " pure " science as well as to the social relations of science.

Scientists are becoming familiar with the application of Marxist ideas to the place of science in society. Some accept it in whole or in part, others fight against it vigorously, and say that they are pursuing pure knowledge for its own sake. But many of them are unaware that Marxism has any bearing on scientific problems considered out of their relation to society, for example to the problems of tautomerism in chemistry or individuality in biology. And certain Marxists are inclined to regard the study of such scientific and philosophical problems as unimportant. Yet they have before them the example of Lenin. In 1908 the Russian Revolution had failed. It was necessary to build up the revolutionary movement afresh. Lenin saw that this could

only be done on a sound theoretical basis.   So he wrote
Materialism and Empirio-criticism.   This involved a
study, not only of philosophers such as Mach and
Pearson, whom he criticised, but of physicists such as
Hertz, J. J. Thomson, and Becquerel, whose discoveries
could be interpreted from a materialistic or an idealistic
point of view.   However, Lenin did not attempt to
cover the whole of science.   He was mainly concerned
with the revolution in physics which was then in progress,
and had little to say on astronomy, geology, chemistry,
or biology.

But thirty years before Lenin, Engels had tried to
discuss the whole of science from a Marxist standpoint.
He had always been a student of science.   Since 1861
he had been in close touch with the chemist Schorlemmer
at Manchester, and had discussed scientific problems
with him and Marx for many years.   In 1871 he came to
London, and started reading scientific books and journals
on a large scale.   He intended to write a great book to
show " that in nature the same dialectical laws of
movement are carried out in the confusion of its countless changes, as also govern the apparent contingency of
events in history."   If this book had been written, it
would have been of immense importance for the development of science.

But apart from political work, other intellectual tasks
lay before Engels.   Dühring had to be answered, and
perhaps *Anti-Dühring*, which covers the whole field
of human knowledge, is a greater book than *Dialectics
of Nature* would have been had Engels completed it.
After Marx's death in 1883 he had the gigantic task
of editing and completing *Capital*, besides which he
wrote *Feuerbach* and *The Origin of the Family*.   So
Dialectics of Nature was never finished.   The manuscript consists of four bundles, all in Engels' handwriting,
save for a number of quotations from Greek philosophers

in that of Marx.   Part of the manuscript is ready for
publication, though, as we shall see, it would almost
certainly have been revised.   Much of it merely consists
of rough notes, which Engels hoped to work up later.
They are often hard to read, and full of abbreviations,
e.g. Mag. for magnet and magnetism.   There are
occasional scribbles and sketches in the margin.   Finally,
although the bulk of the manuscript is in German,
Engels thought equally well in English and French, and
occasionally produced a hybrid sentence, such as
" Wenn Coulomb von particles of electricity spricht,
which repel each other inversely as the square of the
distance, so nimmt Thomson das ruhig hin als bewiesen."
Or "In der heutigen Gesellschaft, dans le méchanisme
civilisé, herrscht duplicité d'action, contrariété de
l'interêt individuel avec le collectif; es ist une queue
universelle des individus contre les masses."   The
translation has been a very difficult task, and the
order of the different parts is somewhat uncertain.

Most of the manuscript seems to have been written
between 1872 and 1882, that is to say it refers to the
science of sixty years ago.   Hence it is often hard to
follow if one does not know the history of the scientific
practice and theory of that time.   The idea of what is
now called the conservation of energy was beginning
to permeate physics, chemistry, and biology.   But it was
still very incompletely realised, and still more incompletely applied.   Words such as " force," " motion,"
and " *vis viva* " were used where we should now speak of
energy.   The essays on " Basic forms of motion,"
" The measure of motion—work," and " Heat " are
largely concerned with the controversies which arose
from incomplete or faulty theories about energy.
They are interesting as showing how ideas on this
subject developed, and how Engels tackled the controversies of his day.   However many of these contro-

versies are now settled.   The expression *vis viva* is no
longer used for double the kinetic energy, and " force "
has acquired a definite meaning in physics.   Engels
would not have published them in their present form,
if only because, in the later essay on tidal friction, he
uses a more modern terminology.   Their interest lies
not so much in their detailed criticism of theories, many
of which have ceased to be of importance, but in showing
how Engels grappled with intellectual problems.   The
essay on electricity " dates " even more.   As a criticism
of Wiedemann's inconsistencies it is interesting, and it
ends with a plea for a closer investigation of the connection between chemical and electrical action, which,
as Engels said, " will lead to important results in both
spheres of investigation."   This prophecy has, of course,
been amply fulfilled.   Arrhenius' ionic theory has transformed   chemistry,   and   Thomson's   electron   theory
has revolutionised physics.   Here again, the manuscript
would certainly have been revised before publication.
In a letter to Marx on November 23rd, 1882, he points
out that Siemens, in his presidential address to the
British Association, has defined a new unit, that of
electric power, the Watt, which is proportional to the
resistance multiplied by the square of the current
whereas the electromotive force is proportional to the
resistance multiplied by the current.   He compares these
with the expressions for momentum and energy, discussed in the essay on " The measure of motion—work,'
and points out that in each case we have simple proportionality (momentum as velocity and electromotive
force as current) when we are not dealing with transformation of one form of energy into another.   But
when the energy is transformed into heat or work the
correct value is found by squaring the velocity or
current.   " So it is a general law of motion which I
was the first to formulate."   We can now see why this

is so. The momentum and the electromotive force, having directions, are reversed when the speed and current are reversed. But the energy remains unaltered. So the speed or the current must come into the formula as the square (or some even power) since $(-x)^2 = x^2$.

In the essay on "Tidal friction," Engels made a serious mistake, or more accurately a mistake which would have been serious had he published it. But I very much doubt whether he would have done so. In the manuscript notes for Anti-Dühring,[1] he supported the view, quite commonly held in the nineteenth century, that we find truths such as mathematical axioms self-evident because our ancestors have been convinced of their validity, while they would not appear self-evident to a Bushman or Australian black. Now this view is almost certainly incorrect, and Engels presumably saw the fallacy, and did not have it printed. I have little doubt that either he or one of his scientific friends such as Schorlemmer would have detected the mistake in the essay on "Tidal friction." But even as a mistake it is interesting, because it is one of the mistakes which lead to a correct result (namely that the day would shorten even if there were no oceans) by incorrect reasoning. Such mistakes have been extremely fruitful in the history of science.

Elsewhere there are statements which are certainly untrue, for example in the sections on stars and Protozoa. But here Engels cannot be blamed for following some of the best astronomers and zoologists of his day. The technical improvement of the telescope and micro-scope has of course led to great increases in our knowledge here in the last sixty years.

On the other hand, Engels' remarks on the differential calculus, though inapplicable to that branch of mathe-matics as now taught, were correct in his own day,

[1] See p. 314.

and for some time after. He points out that it actually developed by contradiction, and is none the worse for that. To-day " rigorous " proofs are given of many of the theorems to which he refers, and some mathematicians claim to have eliminated the contradictions. Actually they have only pushed the contradictions into the background, where they remain in the field of mathematical logic. Not only has every effort to deduce all mathematics from a set of axioms, and rules for applying them, failed, but Gödel has proved that they must fail. So the fact that the calculus can be taught without involving the particular contradictions mentioned by Engels in no way impugns the validity of his dialectical argument.

When all such criticisms have been made, it is astonishing how Engels anticipated the progress of science in the sixty years which have elapsed since he wrote. He certainly did not like the atomic theory of electricity, which held sway from 1900 to 1930, and until it turned out that the electron behaved not only like a particle but like a system of moving waves he might well have been thought to have " backed the wrong horse." His insistence that life is the characteristic mode of behaviour of proteins appeared to be very one-sided to most biochemists, since every cell contains many other complicated organic substances besides proteins. Only in the last four years has it turned out that certain pure proteins do exhibit one of the most essential features of living things, reproducing themselves in a variety of environments.

While we can everywhere study Engels' method of thinking with advantage, I believe that the sections of the book which deal with biology are the most immediately valuable to scientists to-day. This may of course be because as a biologist I can detect subtleties of Engels' thought which I have missed in the physical sections. It may be because biology has undergone less spectacular changes than physics in the last two generations.

In order to help readers to follow the development of science since Engels' time, I have added some notes. A few readers may object to my pointing out that Engels was occasionally wrong. Engels would not have objected. He was well aware that he was not infallible, and that the Labour Movement wants no popes or inspired scriptures. *The Condition of the Working Class in England in* 1844, of which an English translation had been published in America in 1885, was first published in England in 1892. In his preface written after forty-eight years he says :

" I have taken great care not to strike out of the text the many prophecies, amongst others that of an imminent social revolution in England, which my youthful ardour induced me to venture upon. The wonder is, not that a good many of them proved wrong, but that so many of them have proved right."

I think that readers of *Dialectics of Nature* will come to a similar conclusion.

I have not yet mentioned the sections on the history of science. These are among the most brilliant passages in the whole book, but they represent a line of thought which was followed by Marx and Engels in many of their books and which has since been developed by others, so most readers will find them less novel. Finally, there is the delightful essay on " Scientific research into the spirit world." There is a tendency among materialists to neglect the problems here dealt with. It is worth while noticing that Engels did not do so. On the contrary he produced a number of phenomena which were regarded as " occult " and mysterious in his day, and arrived at the same conclusions as most scientific investigators in this field have reached, provided that, like Engels, they brought to their work robust common sense, and also a sense of humour.

It was a great misfortune, not only for Marxism, but for all branches of natural science, that Bernstein, into whose hands the manuscript came when Engels died in 1895, did not publish it. In 1924 he submitted it (or part of it) to Einstein, who, though he did not think it of great interest from the standpoint of modern physics, was on the whole in favour of publication. If, as seems likely, Einstein only saw the essay on electricity, his hesitation can easily be understood, since this deals almost wholly with questions which now seem remote. The manuscript was first edited by Riazanov, and printed in 1927. However, Adoratski's edition of 1935 is more satisfactory, as several passages which made nonsense in the earlier edition have now been deciphered.

Had Engels' method of thinking been more familiar, the transformations of our ideas on physics which have occurred during the last thirty years would have been smoother. Had his remarks on Darwinism been generally known, I for one would have been saved a certain amount of muddled thinking. I therefore welcome wholeheartedly the publication of an English translation of *Dialectics of Nature*, and hope that future generations of scientists will find that it helps them to elasticity of thought.

But it must not be thought that *Dialectics of Nature* is only of interest to scientists. Any educated person, and, above all, anyone who is a student of philosophy, will find much to interest him or her throughout the book, though particularly in Chapters I, II, VII, IX, and X. One reason why Engels was such a great writer is that he was probably the most widely educated man of his day. Not only had he a profound knowledge of economics and history, but he knew enough to discuss the meaning of an obscure Latin phrase concerning

Roman marriage law, or the processes taking place when a piece of impure zinc was dipped into sulphuric acid. And he contrived to accumulate this immense knowledge, not by leading a life of cloistered learning, but while playing an active part in politics, running a business, and even fox-hunting !

He needed this knowledge because dialectical materialism, the philosophy which, along with Marx, he founded, is not merely a philosophy of history, but a philosophy which illuminates all events whatever, from the falling of a stone to a poet's imaginings. And it lays particular emphasis on the inter-connection of all processes, and the artificial character of the distinctions which men have drawn, not merely between vertebrates and invertebrates or liquids and gases, but between the different fields of human knowledge such as economics, history, and natural science.

Chapter II contains an outline of this philosophy in its relation to natural science. A very careful and condensed summary of it is given in Chapter IV of the *History of the C.P.S.U.(B)*, but the main sources for its study are Engels' *Feuerbach* and *Anti-Dühring*, Lenin's *Materialism and Empirio-criticism*, and a number of passages in the works of Marx. Just because it is a living philosophy with innumerable concrete applications its full power and importance can only be gradually understood, when we see it applied to history, science, or whatever field of study interests us most. For this reason a reader whose concern lies primarily in the political or economic field will come back to his main interest a better dialectical materialist, and therefore a clearer-sighted politician or economist, after studying how Engels applied Dialectics to Nature.

At the present moment, clear thinking is vitally necessary if we are to understand the extremely compli- cated situation in which the whole human race, and our own nation in particular, is placed, and to see the way out of it to a better world. A study of Engels will warn us against some of the facile solutions which are put forward to-day, and help us to play an intelligent and courageous part in the great events of our own time.

J. B. S. HALDANE.

November, 1939.

INTRODUCTION

I

MODERN natural science, which alone has achieved an all-round systematic and scientific development, as contrasted with the brilliant natural-philosophical intuitions of antiquity and the extremely important but sporadic discoveries of the Arabs, which for the most part vanished without results —this modern natural science dates, like all more recent history, from that mighty epoch which we Germans term the Reformation, from the national misfortune that overtook us at that time, and which the French term the Renaissance and the Italians the Cinquecento, although it is not fully expressed by any of these names. It is the epoch which had its rise in the last half of the fifteenth century. Royalty, with the support of the burghers of the towns, broke the power of the feudal nobility and established the great monarchies, based essentially on nationality, within which the modern European nations and modern bourgeois society came to development. And while the burghers and nobles were still fighting one another, the peasant war in Germany pointed prophetically to future class struggles, not only by bringing on to the stage the peasants in revolt—that was no longer anything new—but behind them the beginnings of the modern proletariat, with the red flag in their hands and the demand for common ownership of goods on their lips. In the manuscripts saved from the fall of Byzantium, in the antique statues dug out of the ruins of Rome, a new world was revealed to the astonished West, that of ancient Greece : the ghosts of the Middle Ages vanished before its shining forms ; Italy rose to an undreamt-of flowering of art, which seemed like a reflection of classical antiquity and was never attained again. In Italy, France, and Germany a new literature arose, the first modern literature ; shortly afterwards came the classical epochs of English and Spanish literature. The bounds of the old *orbis terrarum* were pierced. Only now for the first time was the world really discovered and the basis laid for subsequent world trade and the transition from handicraft to manufacture, which in its turn formed the starting-point for modern large scale industry. The dictatorship of the Church over men's minds was shattered ; it was directly cast off by the majority of the Germanic peoples, who adopted Protestantism, while among the Latins a cheerful spirit of free thought, taken over from the Arabs and nourished by the newly-discovered Greek philosophy, took root more and more and prepared the way for the materialism of the eighteenth century.

It was the greatest progressive revolution that mankind has so far experienced, a time which called for giants and produced giants—giants in power of thought, passion, and character, in universality and learning. The men who founded the modern rule of the bourgeoisie had anything but bourgeois limitations. On the contrary, the adventurous character of the time inspired them to a greater or less degree. There was hardly any man of importance then living who had not travelled extensively, who did not command four or five languages, who did not shine in a number of fields. Leonardo da Vinci was not only a great painter but also a great mathematician, mechanician, and engineer, to whom the most diverse branches of physics are indebted for important discoveries. Albrecht Dürer was painter, engraver, sculptor, and architect, and in addition invented a system of fortification embodying many of

the ideas that much later were again taken up by Monta-
lembert and the modern German science of fortification.
Machiavelli was statesman, historian, poet, and at the
same time the first notable military author of modern
times. Luther not only cleaned the Augean stable [1] of
the Church but also that of the German language ; he
created modern German prose and composed the text
and melody of that triumphal hymn which became the
Marseillaise of the sixteenth century.[2] The heroes of
that time had not yet come under the servitude of the
division of labour, the restricting effects of which, with
its production of onesidedness, we so often notice in
their successors. But what is especially characteristic
of them is that they almost all pursue their lives and
activities in the midst of the contemporary movements,
in the practical struggle ; they take sides and join
in the fight, one by speaking and writing, another with
the sword, many with both. Hence the fullness and
force of character that makes them complete men.
Men of the study are the exception—either persons of
second or third rank or cautious philistines who do not
want to burn their fingers.

At that time natural science also developed in the
midst of the general revolution and was itself thoroughly
revolutionary ; it had to win in struggle its right of
existence. Side by side with the great Italians from
whom modern philosophy dates, it provided its martyrs
for the stake and the prisons of the Inquisition. And it
is characteristic that Protestants outdid Catholics in
persecuting the free investigation of nature. Calvin
had Servetus burnt at the stake when the latter was
on the point of discovering the circulation of the blood,

[1] Augean stable : one of the mythical labours of the Greek hero
Heracles (Hercules) was the removal of dung from this stable.
[2] *Ein fester Burg ist unser Gott.* (" A safe stronghold our God is
still ") : This hymn has recently been sung on a large scale by protestant congregations in Germany which have not accepted Hitler's
theology.

and indeed he kept him roasting alive during two hours ; for the Inquisition at least it sufficed to have Giordano Bruno simply burnt alive.

The revolutionary act by which natural science declared its independence and, as it were, repeated Luther's burning of the Papal Bull was the publication of the immortal work by which Copernicus, though timidly and, so to speak, only from his deathbed, threw down the gauntlet to ecclesiastical authority in the affairs of nature. The emancipation of natural science from theology dates from this act, although the fighting out of the particular antagonistic claims has dragged out up to our day and in many minds is still far from completion. Thenceforward, however, the development of the sciences proceeded with giant strides, and, it might be said, gained in force in proportion to the square of the distance (in time) from its point of departure. It was as if the world were to be shown that henceforth the reciprocal law of motion would be as valid for the highest product of organic matter, the human mind, as for inorganic substance.

The main work in the first period of natural science that now opened lay in mastering the material immediately at hand. In most fields a start had to be made from the very beginning. Antiquity had bequeathed Euclid and the Ptolemaic solar system ; the Arabs had left behind the decimal notation, the beginnings of algebra, the modern numerals, and alchemy ; the Christian Middle Ages nothing at all. Of necessity, in this situation the most fundamental natural science, the mechanics of terrestrial and heavenly bodies, occupied first place, and alongside of it, as handmaiden to it, the discovery and perfecting of mathematical methods. Great work was achieved here. At the end of the period characterised by Newton and Linnæus we find these branches of science brought to a certain perfection. The basic features of the most essential mathematical methods were established ; analytical geometry by Descartes especially, logarithms by Napier, and the differential and integral calculus by Leibniz and perhaps Newton.[1] The same holds good of the mechanics of rigid bodies, the main laws of which were made clear once for all. Finally in the astronomy of the solar system Kepler discovered the laws of planetary movement and Newton formulated them from the point of view of the general laws of motion of matter. The other branches of natural science were far removed even from this preliminary perfection. Only towards the end of the period did the mechanics of fluid and gaseous bodies receive further treatment. Physics proper had still not gone beyond its first beginnings, with the exception of optics, the exceptional progress of which was due to the practical needs of astronomy. By the phlogistic theory, chemistry for the first time emancipated itself from alchemy. Geology had not yet gone beyond the embryonic stage of mineralogy ; hence palæontology could not yet exist at all. Finally, in the field of biology the essential preoccupation was still with the collection and first sifting of the immense material, not only botanical and zoological but also anatomical and even physiological. There could as yet be hardly any talk of the comparison of the various forms of life, of the investigation of their geographical distribution and their climatic, etc., living conditions. Here only botany and zoology arrived at an approximate completion owing to Linnæus.

[1] There can be little doubt that Newton and Leibniz invented the calculus independently. Here and elsewhere Engels is perhaps over-critical of Newton. It must be remembered that Newton's essentially mechanical outlook on nature had been so brilliantly successful for over a century that it had been accepted as a dogma, and was therefore retarding the progress of science. Now that we can see where Newton went wrong, we can perhaps appreciate his greatness better than was possible when it was absolutely essential to criticise him.

But what especially characterises this period is the elaboration of a peculiar general outlook, in which the central point is the view of the *absolute immutability of nature*. In whatever way nature itself might have come into being, once present it remained as it was as long as it continued to exist. The planets and their satellites, once set in motion by the mysterious "first impulse," circled on and on in their predestined ellipses for all eternity, or at any rate until the end of all things. The stars remained for ever fixed and immovable in their places, keeping one another therein by "universal gravitation." The earth had persisted without alteration from all eternity, or, alternatively, from the first day of its creation. The "five continents" of the present day had always existed, and they had always had the same mountains, valleys, and rivers, the same climate, and the same flora and fauna, except in so far as change or cultivation had taken place at the hand of man. The species of plants and animals had been established once for all when they came into existence ; like continually produced like, and it was already a good deal for Linnæus to have conceded that possibly here and there new species could have arisen by crossing. In contrast to the history of mankind, which develops in time, there was ascribed to the history of nature only an unfolding in space. All change, all development in nature, was denied. Natural science, so revolutionary at the outset, suddenly found itself confronted by an out-and-out conservative nature in which even to-day everything was as it had been at the beginning and in which—to the end of the world or for all eternity— everything would remain as it had been since the beginning.

High as the natural science of the first half of the eighteenth century stood above Greek antiquity in knowledge and even in the sifting of its material, it

stood just as deeply below Greek antiquity in the theoretical mastery of this material, in the general outlook
on nature. For the Greek philosophers the world was
essentially something that had emerged from chaos,
something that had developed, that had come into
being. For the natural scientists of the period that we
are dealing with it was something ossified, something
immutable, and for most of them something that had
been created at one stroke. Science was still deeply
enmeshed in theology. Everywhere it sought and
found its ultimate resort in an impulse from outside
that was not to be explained from nature itself. Even
if attraction, by Newton pompously baptised as " universal gravitation," was conceived as an essential
property of matter, whence comes the unexplained
tangential force which first gives rise to the orbits of
the planets ? How did the innumerable varieties of
animals and plants arise ? And how, above all, did
man arise, since after all it was certain that he was not
present from all eternity ? To such questions natural
science only too frequently answered by making the
creator of all things responsible. Copernicus, at the
beginning of the period, writes a letter renouncing
theology ; Newton closes the period with the postulate
of a divine first impulse. The highest general idea to
which this natural science attained was that of the pur-
posiveness of the arrangements of nature, the shallow
teleology of Wolff, according to which cats were created
to eat mice, mice to be eaten by cats, and the whole
of nature to testify to the wisdom of the creator. It is
to the highest credit of the philosophy of the time that
it did not let itself be led astray by the restricted state of
contemporary natural knowledge, and that—from Spinoza
right to the great French materialists—it insisted on
explaining the world from the world itself and left the
justification in detail to the natural science of the future.

I include the materialists of the eighteenth century in this period because no natural scientific material was available to them other than that above described. Kant's epoch-making work remained a secret to them, and Laplace came long after them. We should not forget that this obsolete outlook on nature, although riddled through and through by the progress of science, dominated the entire first half of the nineteenth century, and in substance is even now still taught in all schools.[1]

The first breach in this petrified outlook on nature was made not by a natural scientist but by a philosopher. In 1755 appeared Kant's *Allgemeine Naturgeschichte und Theorie des Himmels* [*General Natural History and Theory of the Heavens*]. The question of the first impulse was abolished; the earth and the whole solar system appeared as something that had *come into being* in the course of time. If the great majority of the natural scientists had had a little less of the repugnance to thinking that Newton expressed in the warning : " Physics, beware of metaphysics ! ", they would have been compelled from this single brilliant discovery of Kant's to draw conclusions that would have spared them endless deviations and immeasurable amounts of time and labour wasted in false directions.   For Kant's discovery contained the

[1] How tenaciously even in 1861 this view could be held by a man whose scientific achievements had provided highly important material for abolishing it is shown by the following classic words :

" All the arrangements of our solar system, so far as we are capable of comprehending them, aim at preservation of what exists and at unchanging continuance.   Just as since the most ancient times no animal and no plant on the earth has become more perfect or in any way different, just as we find in all organisms only stages *alongside of* one another and not following one another, just as our own race has always remained the same in corporeal respects—so even the greatest diversity in the co-existing heavenly bodies does not justify us in assuming that these forms are merely different stages of development ; it is rather that everything created is equally perfect in itself." (Mädler,* *Popular Astronomy.*   Berlin, 1861, 5th edition, p. 316.)   [*Note by F. Engels.*]

* Mädler, a German astronomer, discussed the motions of the so-called fixed stars.

point of departure for all further progress. If the earth were something that had come into being, then its present geological, geographical, and climatic state, and its plants and animals likewise, must be something that had come into being ; it must have had a history not only of co-existence in space but also of succession in time. If at once further investigations had been resolutely pursued in this direction, natural science would now be considerably further advanced than it is. But what good could come of philosophy ? Kant's work remained without immediate results, until many years later Laplace and Herschel expounded its contents and gave them a deeper foundation, thereby gradually bringing the " nebular hypothesis " [1] into favour. Further discoveries finally brought it victory ; the most important of these were : the proper motion of the fixed stars, the demonstration of a resistant medium in universal space, the proof furnished by spectral analysis of the chemical identity of the matter of the universe and the existence of such glowing nebular masses as Kant had postulated.

It is, however, permissible to doubt whether the majority of natural scientists would so soon have become conscious of the contradiction of a changing earth that bore immutable organisms, had not the dawning conception that nature does not just *exist*, but *comes into being* and *passes away*, derived support from another quarter. Geology arose and pointed out, not only the terrestrial strata formed one after another and deposited

---

[1] This was the hypothesis that the sun and its planets had condensed out of a rotating nebula. It was regarded as plausible for over a century. However, there is now no doubt that the nebulæ are all vastly larger than the solar system, and the spiral nebulæ, from one of which the solar system was thought to have originated, are systems of thousands of millions of stars, like our own Milky Way, but much more distant. The hypothesis was however of immense importance because if first made it likely that the solar system has a history. It may be compared with the ideas of the ancients on biological evolution.

one upon another, but also the shells and skeletons of extinct animals and the trunks, leaves, and fruits of no longer existing plants contained in these strata. It had finally to be acknowledged that not only the earth as a whole but also its present surface and the plants and animals living on it possessed a history in time. At first the acknowledgement occurred reluctantly enough. Cuvier's theory of the revolutions of the earth was revolutionary in phrase and reactionary in substance. In place of a single divine creation, he put a whole series of repeated acts of creation, making the miracle an essential natural agent. Lyell first brought sense into geology by substituting for the sudden revolutions due to the moods of the creator the gradual effects of a slow transformation of the earth.[1]

Lyell's theory was even more incompatible than any of its predecessors with the assumption of constant organic species. Gradual transformation of the earth's surface and of all conditions of life led directly to gradual transformation of the organisms and their adaptation to the changing environment, to the mutability of species. But tradition is a power not only in the Catholic Church but also in natural science. For years, Lyell himself did not see the contradiction, and his pupils still less. This is only to be explained by the division of labour that had meanwhile become dominant in natural science, which more or less restricted each person to his special sphere, there being only a few whom it did not rob of a comprehensive view.

Meanwhile physics had made mighty advances, the results of which were summed up almost simultaneously by three different persons in the year 1842, an epoch-

---

[1] The defect of Lyell's view—at least in its first form—lay in conceiving the forces at work on the earth as constant, both in quality and quantity. The cooling of the earth does not exist for him ; the earth does not develop in a definite direction but merely changes in an inconsequent fortuitous manner. [*Note by F. Engels.*]

making year for this branch of natural investigation. Mayer in Heilbronn and Joule in Manchester demonstrated the transformation of heat into mechanical energy [1] and of mechanical energy into heat. The determination of the mechanical equivalent of heat put this result beyond question. Simultaneously, by simply working up the separate physical results already arrived at, Grove—not a natural scientist by profession, but an English lawyer—proved that all so-called physical energy, mechanical energy, heat, light, electricity, magnetism, indeed even so-called chemical energy, become transformed into one another under definite conditions without any loss of energy occurring, and so proved *post factum* along physical lines Descartes' principle that the quantity of motion present in the world is constant. With that the special physical energies, the as it were immutable " species " of physics, were resolved into variously differentiated forms of the motion of matter, convertible into one another according to definite laws. The fortuitousness of the existence of a number of physical energies was abolished from science by the proof of their interconnections and transitions. Physics, like astronomy before it, had arrived at a result that necessarily pointed to the eternal cycle of matter in motion as the ultimate reality.

The wonderfully rapid development of chemistry, since Lavoisier, and especially since Dalton, attacked the old ideas of nature from another aspect. The preparation by inorganic means of compounds that hitherto had been produced only in the living organism proved

[1] Throughout this paragraph the German word " Kraft " has been translated " energy." Joule and other contemporaries used the word " force " where we should now use " energy." We shall see later (p. 49), that Engels objected to the use of the word " Kraft " or force for energy. At one time he preferred " motion," but in his later writings he used the term " energy " as almost all modern writers do. The rendering here makes Engels' meaning clearer than if the ambiguous word " force " had been used.

that the laws of chemistry have the same validity for organic as for inorganic bodies, and to a large extent bridged the gulf between inorganic and organic nature, a gulf that even Kant regarded as for ever impassable.

Finally, in the sphere of biological research also the scientific journeys and expeditions that had been systematically organised since the middle of the previous century, the more thorough exploration of the European colonies in all parts of the world by specialists living there, and further the progress of palæontology, anatomy, and physiology in general, particularly since the systematic use of the microscope and the discovery of the cell, had accumulated so much material that the application of the comparative method became possible and at the same time indispensable. On the one hand the conditions of life of the various floras and faunas were determined by means of comparative physical geography; on the other hand the various organisms were compared with one another according to their homologous organs, and this not only in the adult condition but at all stages of development. The more deeply and exactly this research was carried on, the more did the rigid system of an immutable, fixed organic nature crumble away at its touch. Not only did the separate species of plants and animals become more and more inextricably intermingled, but animals turned up, such as *Amphioxus* [1] and *Lepidosiren*, [2] that made a mockery of all previous classification, and finally organisms were encountered of which it was not possible to say whether they belonged to the plant or animal kingdom. More and more the gaps in the palæontological record were filled up, compelling even the most reluctant to acknowledge the striking parallelism between the evolutionary history of

[1] *Amphioxus*. A headless marine animal with some of the characteristics of a fish, but much more primitive.
[2] *Lepidosiren*. One of the lungfish which can breathe air for months on end.

the organic world as a whole and that of the individual organism, the Ariadne's thread that was to lead the way out of the labyrinth in which botany and zoology appeared to have become more and more deeply lost. It was characteristic that, almost simultaneously with Kant's attack on the eternity of the solar system, C. F. Wolff in 1759 launched the first attack on the fixity of species and proclaimed the theory of descent. But what in his case was still only a brilliant anticipation took firm shape in the hands of Oken, Lamarck, Baer, and was victoriously carried through by Darwin in 1859, exactly a hundred years later. Almost simultaneously it was established that protoplasm and the cell, which had already been shown to be the ultimate morphological constituents of all organisms, occurred independently as the lowest forms of organic life. This not only reduced the gulf between inorganic and organic nature to a minimum but removed one of the most essential difficulties that had previously stood in the way of the theory of descent of organisms. The new conception of nature was complete in its main features ; all rigidity was dissolved, all fixity dissipated, all particularity that had been regarded as eternal became transient, the whole of nature shown as moving in eternal flux and cyclical course.

Thus we have once again returned to the point of view of the great founders of Greek philosophy, the view that the whole of nature, from the smallest element to the greatest, from grains of sand to suns, from protista [1] to men, has its existence in eternal coming into being and passing away, in eeaseless flux, in unresting motion and change, only with the essential difference that what for the Greeks was a brilliant intuition, is in our case the result of strictly scientific

[1] *Protista.* Single-celled animals and plants such as Paramœcium, Amœba, Bacillus.

research in accordance with experience, and hence also it emerges in a much more definite and clear form. It is true that the empirical proof of this motion is not wholly free from gaps, but these are insignificant in comparison with what has already been firmly established, and with each year they become more and more filled up. And how could the proof in detail be otherwise than defective when one bears in mind that the most essential branches of science—trans-planetary astronomy, chemistry, geology—have a scientific existence of barely a hundred years, and the comparative method in physiology one of barely fifty years, and that the basic form of almost all organic development, the cell, is a discovery not yet forty years old ?

The innumerable suns and solar systems of our island universe,[1] bounded by the outermost stellar rings of the Milky Way, developed from swirling, glowing masses of vapour, the laws of motion of which will perhaps be disclosed after the observations of some centuries have given us an insight into the proper motion of the stars. Obviously, this development did not proceed everywhere at the same rate. Recognition of the existence of dark bodies, not merely planetary in nature, hence extinct suns in our stellar system, more and more forces itself on astronomy (Mädler); on the other hand (according to Secchi) a part of the vaporous nebular patches belong to our stellar system as suns not yet fully formed, whereby it is not excluded that other nebulæ, as Mädler maintains, are distant independent island universes, the relative stage of development of which must be determined by the spectroscope.

[1] This refers to the system of stars of which the sun is one, and the Milky Way represents the densest portions. Mädler was right in maintaining that many of the other bodies then described as nebulæ were similar masses of stars. His view that there are extinct suns is more doubtful. Nor is it clear that the gaseous nebulæ are likely to condense into suns.

How a solar system develops from an individual nebular mass has been shown in detail by Laplace in a manner still unsurpassed ; subsequent science has more and more confirmed him.[1]

On the separate bodies so formed—suns as well as planets and satellites—the form of motion of matter at first prevailing is that which we call heat. There can be no question of chemical compounds of the elements even at a temperature like that still possessed by the sun ; the extent to which heat is transformed into electricity or magnetism [2] under such conditions, continued solar observations will show; it is already as good as proved that the mechanical motion taking place in the sun arises solely from the conflict of heat with gravity.

The smaller the individual bodies, the quicker they cool down, the satellites, asteroids, and meteors first of all, just as our moon has long been extinct. The planets cool more slowly, the central body slowest of all.

With progressive cooling the interplay of the physical forms of motion which become transformed into one another comes more and more to the forefront until finally a point is reached from when on chemical affinity begins to make itself felt, the previously chemically indifferent elements become differentiated chemically one after another, obtain chemical properties, and enter into combination with one another. These compounds change continually with the decreasing temperature, which affects differently not only each element but also each separate compound of the elements, changing also with the consequent passage of part of the gaseous matter first to the liquid and then the solid state, and with the new conditions thus created.

[1] Laplace's theory is fairly certainly incorrect.
[2] Huge magnetic fields have been discovered in the sunspots, and it is also known that the matter shot out in solar prominences is electrically charged. Both these facts were unsuspected by most, if not all, astronomers when Engels wrote.

The period when the planet has a firm shell and accumulations of water on its surface coincides with that when its intrinsic heat diminishes more and more in comparison to the heat emitted to it from the central body. Its atmosphere becomes the arena of meteorological phenomena in the sense in which we now understand the word ; its surface becomes the arena of geological changes in which the deposits resulting from atmospheric precipitation become of ever greater importance in comparison to the slowly decreasing external effects of the hot fluid interior.

If, finally, the temperature becomes so far equalised that over a considerable portion of the surface at least it does not exceed the limits within which protein [1] is capable of life, then, if other chemical conditions are favourable, living protoplasm is formed. What these conditions are, we do not yet know, which is not to be wondered at since so far not even the chemical formula of protein has been established—we do not even know how many chemically different protein bodies there are—and since it is only about ten years ago that the fact became known that completely structureless protein [2] exercises all the essential functions of life, digestion, excretion, movement, contraction, reaction to stimuli, and reproduction.

---

[1] Throughout this book Engels' word "*Eiweiss*" is translated as "*protein*." The word "*albumen*," which has been used in the translation of some of Engels' other works, is now applied to one group of the proteins only. The chemical formulæ of a few proteins were first discovered with fair accuracy by Bergmann, a German-Jewish refugee in New York, in 1936. However, the order in which their constituents are arranged is still incompletely known. There are probably many millions of different proteins.

[2] *Structureless protein : Bathybius Haeckeli*, which was supposed to be an organism composed of a mere mass of structureless protein, proved to be an artefact, that is to say not a natural object, but one produced by the chemicals intended to preserve it. However Engels was fundamentally right. Some of the " viruses," that is to say the smallest agents of disease, are simply large protein molecules, as first shown by Stanley in 1936. They do not appear to exercise all the functions of life, but only some of them.

Thousands of years may have passed before the conditions arose in which the next advance could take place and this formless protein produce the first cell by formation of nucleus and cell membrane. But this first cell also provided the foundation for the morphological development of the whole organic world; the first to develop, as it is permissible to assume from the whole analogy of the palæontological record, were innumerable species of non-cellular and cellular protista, of which *Eozoon canadense* [1] alone has come down to us, and of which some were gradually differentiated into the first plants and others into the first animals. And from the first animals were developed, essentially by further differentiation, the numerous classes, orders, families, genera, and species of animals; and finally mammals, the form in which the nervous system attains its fullest development; and among these again finally that mammal in which nature attains consciousness of itself—man.

Man too arises by differentiation. Not only individually, by differentiation from a single egg cell to the most complicated organism that nature produces—no, also historically. When after thousands of years [2] of struggle the differentiation of hand from foot, and erect gait, were finally established, man became distinct from the monkey and the basis was laid for the development of articulate speech and the mighty development of the brain that has since made the gulf between man and monkey an unbridgeable one. The specialisation of the hand—this implies the *tool*, and the tool implies specific human activity, the transforming reaction of man on nature, production. Animals in the narrower sense also have tools, but only as limbs of their bodies:

[1] *Eozoon canadense* is almost certainly not an organic product. Nevertheless there is every reason to believe in the essential truth of this paragraph.
[2] The geological time-scale is longer than was believed fifty years ago. " Millions of years " would be more correct.

the ant, the bee, the beaver; animals also produce, but their productive effect on surrounding nature in relation to the latter amounts to nothing at all. Man alone has succeeded in impressing his stamp on nature, not only by shifting the plant and animal world from one place to another, but also by so altering the aspect and climate of his dwelling place, and even the plants and animals themselves, that the consequences of his activity can disappear only with the general extinction of the terrestrial globe. And he has accomplished this primarily and essentially by means of *the hand*. Even the steam engine, so far his most powerful tool for the transformation of nature, depends, because it is a tool, in the last resort on the hand. But step by step with the development of the hand went that of the brain; first of all consciousness of the conditions for separate practically useful actions, and later, among the more favoured peoples and arising from the preceding, insight into the natural laws governing them. And with the rapidly growing knowledge of the laws of nature the means for reacting on nature also grew; the hand alone would never have achieved the steam engine if the brain of man had not attained a correlative development with it, and parallel to it, and partly owing to it.

With men we enter *history*. Animals also have a history, that of their derivation and gradual evolution to their present position. This history, however, is made for them, and in so far as they themselves take part in it, this occurs without their knowledge or desire. On the other hand, the more that human beings become removed from animals in the narrower sense of the word, the more they make their own history consciously, the less becomes the influence of unforeseen effects and uncontrolled forces on this history, and the more accurately does the historical result correspond to the aim laid down in advance. If, however, we apply this

INTRODUCTION   19

measure to human history, to that of even the most
developed peoples of the present day, we find that there
still exists here a colossal disproportion between the
proposed aims and the results arrived at, that unforeseen
effects predominate, and that the uncontrolled forces
are far more powerful than those set into motion
according to plan. And this cannot be otherwise as
long as the most essential historical activity of men, the
one which has raised them from bestiality to humanity
and which forms the material foundation of all their
other activities, namely the production of their requirements of life, that is to-day social production, is above
all subject to the interplay of unintended effects from
uncontrolled forces and achieves its desired end only
by way of exception and, much more frequently, the
exact opposite. In the most advanced industrial
countries we have subdued the forces of nature and
pressed them into the service of mankind ; we have
thereby infinitely multiplied production, so that a child
now produces more than a hundred adults previously
did. And what is the result ? Increasing overwork
and increasing misery of the masses, and every ten years
a great collapse. Darwin did not know what a bitter
satire he wrote on mankind, and especially on his
countrymen, when he showed that free competition, the
struggle for existence, which the economists celebrate
as the highest historical achievement, is the normal state
of the *animal kingdom*. Only conscious organisation of
social production, in which production and distribution
are carried on in a planned way, can lift mankind
above the rest of the animal world as regards the social
aspect, in the same way that production in general has
done this for men in their aspect as species. Historical
evolution makes such an organisation daily more indispensable, but also with every day more possible.
From it will date a new epoch of history, in which man-

kind itself, and with mankind all branches of its activity, and especially natural science, will experience an advance that will put everything preceding it in the deepest shade.

Nevertheless, "all that comes into being deserves to perish." Millions of years may elapse, hundreds of thousands of generations be born and die, but inexorably the time will come when the declining warmth of the sun [1] will no longer suffice to melt the ice thrusting itself forward from the poles; when the human race, crowding more and more about the equator, will finally no longer find even there enough heat for life; when gradually even the last trace of organic life will vanish; and the earth, an extinct frozen globe like the moon, will circle in deepest darkness and in an ever narrower orbit about the equally extinct sun, and at last fall into it. Other planets will have preceded it, others will follow it; instead of the bright, warm solar system with its harmonious arrangement of members, only a cold, dead sphere will still pursue its lonely path through universal space. And what will happen to our solar system will happen sooner or later to all the other systems of our island universe; it will happen to all the other innumerable island universes, even to those the light of which will never reach the earth while there is a living human eye to receive it.

And when such a solar system has completed its life history and succumbs to the fate of all that is finite, death, what then? Will the sun's corpse roll on for all eternity through infinite space, and all the once infinitely diverse, differentiated natural forces pass for ever into

[1] Until quite recently these rather gloomy conclusions appeared inevitable, even if the time-scale proved to be vastly longer than was supposed. But in 1936–1938 Milne and Dirac independently arrived at the conclusion that the laws of nature themselves evolve, and in particular (according to Milne) that chemical changes are speeded up (at the rate of about one two-thousand-millionth part per year) in relation to physical changes. If so it is at least conceivable that this process may be rapid enough to compensate for the cooling of the stars, and that life may never become impossible.

one single form of motion, attraction ? " Or "—as
Secchi asks (p. 810)—" do forces exist in nature which
can re-convert the dead system into its original state
of an incandescent nebula and re-awake it to new life ?
We do not know."

At all events we do not know in the sense that we
know that $2 \times 2 = 4$, or that the attraction of matter
increases and decreases according to the square of the
distance. In theoretical natural science, however,
which as far as possible builds up its view of nature into
a harmonious whole, and without which nowadays
even the most thoughtless empiricist cannot get anywhere, we have very often to reckon with incompletely
known magnitudes; and logical consistency of thought
must at all times help to get over defective knowledge.
Modern natural science has had to take over from
philosophy the principle of the indestructibility of
motion ; it cannot any longer exist without this principle.
But the motion of matter is not merely crude mechanical
motion, mere change of place, it is heat and light, electric
and magnetic stress, chemical combination and dissociation, life and, finally, consciousness. To say that
matter during the whole unlimited time of its existence
has only once, and for what is an infinitesimally short
period in comparison to its eternity, found itself able
to differentiate its motion and thereby to unfold the
whole wealth of this motion, and that before and after
this remains restricted for eternity to mere change of
place—this is equivalent to maintaining that matter is
mortal and motion transitory. The indestructibility of
motion cannot be merely quantitative, it must also
be conceived qualitatively ; matter whose purely
mechanical change of place includes indeed the possibility
under favourable conditions of being transformed into
heat, electricity, chemical action, or life, but which is not
capable of producing these conditions from out of itself,

such matter has *forfeited motion* ; motion which has lost the capacity of being transformed into the various forms appropriate to it may indeed still have *dynamis* but no longer *energeia*,[1] and so has become partially destroyed. Both, however, are unthinkable.

This much is certain : there was a time when the matter of our island universe had *transformed* a quantity of motion—of what kind we do not yet know—into heat, such that there could be developed from it the solar systems appertaining to (according to Mädler) at least twenty million stars, the gradual extinction of which is likewise certain. How did this transformation take place ? We know just as little as Father Secchi knows whether the future *caput mortuum* of our solar system will once again be converted into the raw material of a new solar system. But here either we must have recourse to a creator, or we are forced to the conclusion that the incandescent raw material for the solar system of our universe was produced in a natural way by transformations of motion which are *by nature inherent* in moving matter, and the conditions of which therefore also must be reproduced by matter, even if only after millions and millions of years and more or less by chance but with the necessity that is also inherent in chance.

The possibility of such a transformation is more and more being conceded. The view is being arrived at that the heavenly bodies are ultimately destined to fall into one another, and one even calculates the amount of heat which must be developed on such collisions. The sudden flaring up of new stars, and the equally sudden increase in brightness of familiar ones, of which we are informed by astronomy, is most easily explained [2] by

---

[1] " Dynamis " and " energeia " are Greek words used by Aristotle. They can roughly be translated as " power " and " activity."

[2] The flaring up of new stars is now generally explained not by collision, but by an internal crisis in the star, in fact in a more dialectical manner.

such collisions. Not only does our group of planets move about the sun, and our sun within our island universe, but our whole island universe also moves in space in temporary, relative equilibrium with the other island universes, for even the relative equilibrium of freely moving bodies can only exist where the motion is reciprocally determined ; and it is assumed by many that the temperature in space is not everywhere the same. Finally, we know that, with the exception of an infinitesimal portion, the heat of the innumerable suns of our island universe vanishes into space and fails to raise the temperature of space even by a millionth of a degree centigrade.[1] What becomes of all this enormous quantity of heat ? Is it for ever dissipated in the attempt to heat universal space, has it ceased to exist practically, and does it only continue to exist theoretically, in the fact that universal space has become warmer by a decimal fraction of a degree beginning with ten or more noughts ? The indestructibility of motion forbids such an assumption, but it allows the possibility that by the successive falling into one another of the bodies of the universe all existing mechanical motion will be converted into heat and the latter radiated into space, so that in spite of all " indestructibility of force " all motion in general would have ceased. (Incidentally it is seen here how inaccurate is the term " indestructibility of force "[2] instead of " indestructibility of motion.") Hence we arrive at the conclusion that in some way, which it will later be the task of scientific research to demonstrate, the heat radiated into space must be able to become transformed into another form of motion, in which it can once more be stored up and rendered active. Thereby the chief difficulty in the way of the

[1] Actually the temperature of dust particles in the space between the galaxies is probably several degrees above absolute zero.
[2] Engels rightly protests against the use of the same word " Kraft " for " force " and " energy."

reconversion of extinct suns into incandescent vapour disappears.

For the rest, the eternally repeated succession of worlds in infinite time is only the logical complement to the co-existence of innumerable worlds in infinite space—a principle the necessity of which has forced itself even on the anti-theoretical Yankee brain of Draper.[1]

It is an eternal cycle [2] in which matter moves, a cycle that certainly only completes its orbit in periods of time for which our terrestrial year is no adequate measure, a cycle in which the time of highest development, the time of organic life and still more that of the life of beings conscious of nature and of themselves, is just as narrowly restricted as the space in which life and self-consciousness come into operation ; a cycle in which every finite mode of existence of matter, whether it be sun or nebular vapour, single animal or genus of animals, chemical combination or dissociation, is equally transient, and wherein nothing is eternal but eternally changing, eternally moving matter and the laws according to which it moves and changes. But however often, and however relentlessly, this cycle is completed in time and space, however many millions of suns and earths may arise and pass away, however long it may last before the conditions for organic life

[1] " The multiplicity of worlds in infinite space leads to the conception of a succession of worlds in infinite time." J. W. Draper, *History of the Intellectual Development of Europe*, 1864. Vol. 2, p. 325. [*Note by F. Engels.*]

[2] At present physicists are divided on this question. A few take Engels' view that the universe goes through cyclical changes, entropy being somehow diminished by processes at present unknown (*e.g.* formation of matter from radiation in interstellar space). Others think as Clausius (see p. 157) did, that it will run down. But there is a third possibility. As pointed out above, the work of Milne suggests that the universe as a whole has a history, though probably an infinite one both in the past and the future. It is almost certain that Engels would have welcomed this idea, although he here admits the eternity of the laws according to which matter moves and changes. But p. 202 shows how close Engels came to Milne's point of view.

develop, however innumerable the organic beings that have to arise and to pass away before animals with a brain capable of thought are developed from their midst, and for a short span of time find conditions suitable for life, only to be exterminated later without mercy, we have the certainty that matter remains eternally the same in all its transformations, that none of its attributes can ever be lost, and therefore, also, that with the same iron necessity that it will exterminate on the earth its highest creation, the thinking mind, it must somewhere else and at another time again produce it.

II

DIALECTICS

(The general nature of dialectics to be developed as the science of interconnections, in contrast to metaphysics.)

It is, therefore, from the history of nature and human society that the laws of dialectics are abstracted. For they are nothing but the most general laws of these two aspects of historical development, as well as of thought itself. And indeed they can be reduced in the main to three :

The law of the transformation of quantity into quality and *vice versa* ;

The law of the interpenetration of opposites ;

The law of the negation of the negation.

All three are developed by Hegel in his idealist fashion as mere laws of *thought* : the first, in the first part of his *Logic*, in the *Doctrine of Being* ; the second fills the whole of the second and by far the most important part of his *Logic*, the *Doctrine of Essence* ; finally the third figures as the fundamental law for the construction of the whole system. The mistake lies in the fact that these laws are foisted on nature and history as laws of thought, and not deduced from them. This is the source of the whole forced and often outrageous treatment ; the universe, willy-nilly, is made out to be arranged in accordance with a system of thought which itself is only the product of a definite stage of evolution of human thought. If we turn the thing round, then everything becomes simple, and the dialectical laws that look so extremely mysterious in idealist philosophy at once become simple and clear as noonday.

Moreover, anyone who is even only slightly acquainted with his Hegel will be aware that in hundreds of passages Hegel is capable of giving the most striking individual illustrations from nature and history of the dialectical laws.

We are not concerned here with writing a handbook of dialectics, but only with showing that the dialectical laws are really laws of development of nature, and therefore are valid also for theoretical natural science. Hence we cannot go into the inner interconnection of these laws with one another.

1. The law of the transformation of quantity into quality and *vice versa*. For our purpose, we could express this by saying that in nature, in a manner exactly fixed for each individual case, qualitative changes can only occur by the quantitative addition or subtraction of matter or motion (so-called energy).[1]

All qualitative differences in nature rest on differences of chemical composition or on different quantities or forms of motion (energy) or, as is almost always the case, on both. Hence it is impossible to alter the quality of a body without addition or subtraction of matter or motion, *i.e.* without quantitative alteration of the body concerned. In this form, therefore, Hegel's mysterious principle appears not only quite rational but even rather obvious.

It is surely hardly necessary to point out that the various allotropic [2] and aggregational states of bodies,

[1] This section was presumably written at a later date than the first. The word *energy* was being used to supersede *force* and *motion* in so far as they measured capacity for doing work.

[2] A substance is said to be *allotropic* when its molecules or atoms can be arranged in different ways so as to give substances with different properties. Thus diamond and graphite are allotropic forms of carbon. The fact that they have different energy contents was foreseen by Engels, but only proved after his death.

because they depend on various groupings of the molecules, depend on greater or lesser quantities of motion communicated to the bodies.

But what is the position in regard to change of form of motion, or so-called energy ? If we change heat into mechanical motion or *vice versa*, is not the quality altered while the quantity remains the same ? Quite correct. But it is with change of form of motion as with Heine's vices ; anyone can be virtuous by himself, for vices two are always necessary. Change of form of motion is always a process that takes place between at least two bodies, of which one loses a definite quantity of motion of one quality (*e.g.* heat), while the other gains a corresponding quantity of motion of another quality (mechanical motion, electricity, chemical decomposition). Here, therefore, quantity and quality mutually correspond to each other. So far it has not been found possible to convert motion from one form to another inside a single isolated body.

We are concerned here in the first place with non-living bodies ; the same law holds for living bodies, but it operates under very complex conditions and at present quantitative measurement is still often impossible for us.[1]

If we imagine any non-living body cut up into smaller and smaller portions, at first no qualitative change occurs. But this has a limit : if we succeed, as by evaporation, in obtaining the separate molecules in the free state, then it is true that we can usually divide these still further, yet only with a complete change of quality. The molecule is decomposed into its separate atoms, which have quite different properties from those of the molecule. In the case of molecules composed of various chemical elements, atoms or molecules of these elements

[1] Engels' view has been completely confirmed by very careful measurements.

themselves make their appearance in the place of the compound molecule; in the case of molecules of elements, the free atoms appear, which exert quite distinct qualitative effects: the free atoms of nascent oxygen are easily able to effect what the atoms of atmospheric oxygen, bound together in the molecule, can never achieve.

But the molecule is also qualitatively different from the mass of the body to which it belongs. It can carry out movements independently of this mass and while the latter remains apparently at rest, *e.g.* heat oscillations; by means of a change of position and of connection with neighbouring molecules it can change the body into an allotrope or a different state of aggregation.

Thus we see that the purely quantitative operation of division has a limit at which it becomes transformed into a qualitative difference: the mass consists solely of molecules, but it is something essentially different from the molecule, just as the latter is different from the atom. It is this difference that is the basis for the separation of mechanics, as the science of heavenly and terrestrial masses, from physics, as the mechanics of the molecule, and from chemistry, as the physics of the atom.

In mechanics, no qualities occur; at most, states such as equilibrium, motion, potential energy, which all depend on measurable transference of motion and are themselves capable of quantitative expression. Hence, in so far as qualitative change takes place here, it is determined by a corresponding quantitative change.

In physics, bodies are treated as chemically unalterable or indifferent; we have to do with changes of their molecular states and with the change of form of the motion which in all cases, at least on one of the two sides, brings the molecule into play. Here every change is a transformation of quantity into quality, a consequence of the quantitative change of the quantity of motion of one form or another that is inherent in the body or communicated to it. " Thus, for instance, the temperature of water is first of all indifferent in relation to its state as a liquid ; but by increasing or decreasing the temperature of liquid water a point is reached at which this state of cohesion alters and the water becomes transformed on the one side into steam and on the other into ice." (Hegel, *Encyclopedia*, Collected Works, VI, p. 217.) Similarly, a definite minimum current strength is required to cause the platinum wire of an electric incandescent lamp to glow ; and every metal has its temperature of incandescence and fusion, every liquid its definite freezing and boiling point at a given pressure—in so far as our means allow us to produce the temperature required ; finally also every gas has its critical point at which it can be liquefied by pressure and cooling. In short, the so-called physical constants are for the most part nothing but designations of the nodal points at which quantitative addition or subtraction of motion produces qualitative alteration in the state of the body concerned, at which, therefore, quantity is transformed into quality.[1]

The sphere, however, in which the law of nature discovered by Hegel celebrates its most important triumphs is that of chemistry. Chemistry can be termed the science of the qualitative changes of bodies as a result of changed quantitative composition. That was already known to Hegel himself (*Logic*, Collected Works, III, p. 433).[2] As in the case of oxygen :

[1] Here, as so often, Engels was far in advance of his time. It was obvious fifty years ago that the melting point of a substance was a nodal point. But we now know that its colour also represents a series of nodal points. As the frequency of light increases from red to violet, we come to a series of frequencies which are just able to set the molecules spinning or vibrating in a particular manner. Light of these frequencies is therefore absorbed. And the colour of a substance is simply the expression of its capacity for absorbing lights of different frequencies. Other examples could be given.

[2] See Appendix II, p. 331.

if three atoms unite into a molecule, instead of the usual two, we get ozone, a body which is very considerably different from ordinary oxygen in its odour and reactions. Again, one can take the various proportions in which oxygen combines with nitrogen or sulphur, each of which produces a substance qualitatively different from any of the others! How different laughing gas (nitrogen monoxide N_2O) is from nitric anhydride (nitrogen pentoxide, N_2O_5)! The first is a gas, the second at ordinary temperatures a solid crystalline substance. And yet the whole difference in composition is that the second contains five times as much oxygen as the first, and between the two of them are three more oxides of nitrogen (NO, N_2O_3, NO_2), each of which is qualitatively different from the first two and from each other.

This is seen still more strikingly in the homologous series of carbon compounds, especially in the simpler hydrocarbons. Of the normal paraffins, the lowest is methane, CH_4; here the four linkages of the carbon atom are saturated by four atoms of hydrogen. The second, ethane, C_2H_6, has two atoms of carbon joined together and the six free linkages are saturated by six atoms of hydrogen. And so it goes on, with C_3H_8, C_4H_{10}, etc., according to the algebraic formula C_nH_{2n+2}, so that by each addition of CH_2 a body is formed that is qualitatively distinct from the preceding one. The three lowest members of the series are gases, the highest known,[1] hexadecane, $C_{16}H_{34}$, is a solid body with a boiling point of 270° C. Exactly the same holds good for the series of primary alcohols with formula $C_nH_{2n+2}O$, derived (theoretically) from the paraffins, and the series of monobasic fatty acids (formula $C_nH_{2n}O_2$). What qualitative difference can be caused by the quantitative

[1] Since Engels' time many more members of the series have been made.

addition of C_3H_6 is taught by experience if we consume
ethyl alcohol, C_2H_6O, in any drinkable form without
addition of other alcohols, and on another occasion
take the same ethyl alcohol but with a slight addition of
amyl alcohol, $C_5H_{12}O$, which forms the main constituent
of the notorious fusel oil. One's head will certainly
be aware of it the next morning, much to its detriment;
so that one could even say that the intoxication, and
subsequent "morning after" feeling, is also quantity
transformed into quality, on the one hand of ethyl
alcohol and on the other hand of this added C_3H_6.

In these series we encounter the Hegelian law in yet
another form. The lower members permit only of a
single mutual arrangement of the atoms. If, however,
the number of atoms united into a molecule attains a
size definitely fixed for each series, the grouping of the
atoms in the molecule can take place in more than one
way; so that two or more isomeric substances can be
formed, having equal numbers of C, H, and O atoms in
the molecule but nevertheless qualitatively distinct
from one another. We can even calculate how many
such isomers are possible for each member of the series.
Thus, in the paraffin series, for C_4H_{10} there are two,
for C_5H_{12} there are three; among the higher members
the number of possible isomers mounts very rapidly.
Hence once again it is the quantitative number of atoms
in the molecule that determines the possibility and, in
so far as it has been proved, also the actual existence of
such qualitatively distinct isomers.

Still more. From the analogy of the substances with
which we are acquainted in each of these series, we can
draw conclusions as to the physical properties of the
still unknown members of the series and, at least for the
members immediately following the known ones, predict
their properties, boiling point, etc., with fair certainty.

Finally, the Hegelian law is valid not only for compound substances but also for the chemical elements themselves. We now know that " the chemical properties of the elements are a periodic function of their atomic weights " (Roscoe-Schorlemmer, *Complete Text-Book of Chemistry*, II, p. 823), and that, therefore, their quality is determined by the quantity of their atomic weight. And the test of this has been brilliantly carried out. Mendeleyev proved that various gaps occur in the series of related elements arranged according to atomic weights indicating that here new elements remain to be discovered. He described in advance the general chemical properties of one of these unknown elements, which he termed eka-aluminium, because it follows after aluminium in the series beginning with the latter, and he predicted its approximate specific and atomic weight as well as its atomic volume. A few years later, Lecoq de Boisbaudran actually discovered this element, and Mendeleyev's predictions fitted with only very slight discrepancies. Eka-aluminium was realised in gallium (*ibid.*, p. 828). By means of the—unconscious—application of Hegel's law of the transformation of quantity into quality, Mendeleyev achieved a scientific feat which it is not too bold to put on a par with that of Leverrier in calculating the orbit of the still unknown planet Neptune.

In biology, as in the history of human society, the same law holds good at every step, but we prefer to dwell here on examples from the exact sciences, since here the quantities are accurately measurable and traceable.

Probably the same gentlemen who up to now have decried the transformation of quantity into quality as mysticism and incomprehensible transcendentalism will now declare that it is indeed something quite self-evident, trivial, and commonplace, which they have long employed, and so they have been taught nothing new.

But to have formulated for the first time in its universally valid form a general law of development of nature, society, and thought, will always remain an act of historic importance. And if these gentlemen have for years caused quantity and quality to be transformed into one another, without knowing what they did, then they will have to console themselves with Molière's Monsieur Jourdain who had spoken prose all his life without having the slightest inkling of it.

III

BASIC FORMS OF MOTION

MOTION in the most general sense, conceived as the mode of existence, the inherent attribute, of matter, comprehends all changes and processes occurring in the universe, from mere change of place right to thinking. The investigation of the nature of motion had as a matter of course to start from the lowest, simplest forms of this motion and to learn to grasp these before it could achieve anything in the way of explanation of the higher and more complicated forms. Hence, in the historical evolution of the natural sciences we see how first of all the theory of simplest change of place, the mechanics of heavenly bodies and terrestrial masses, was developed ; it was followed by the theory of molecular motion, physics, and immediately afterwards, almost alongside of it and in some places in advance of it, the science of the motion of atoms, chemistry. Only after these different branches of the knowledge of the forms of motion governing non-living nature had attained a high degree of development could the explanation of the processes of motion represented by the life process be successfully tackled. This advanced in proportion with the progress of mechanics, physics, and chemistry. Consequently, while mechanics has for a fairly long time already been able adequately to refer the effects in the animal body of the bony levers set into motion by muscular contraction to the laws that prevail also in non-living nature, the physico-chemical establishment of the other phenomena of life is still pretty much at the beginning of its course.[1]   Hence, in investigating here the nature of motion, we are compelled to leave the organic forms of motion out of account.   We are compelled to restrict ourselves—in accordance with the state of science—to the forms of motion of non-living nature.

All motion is bound up with some change of place, whether it be change of place of heavenly bodies, terrestrial masses, molecules, atoms, or ether particles. The higher the form of motion, the smaller this change of place. It in no way exhausts the nature of the motion concerned, but it is inseparable from the motion. It, therefore, has to be investigated before anything else.

The whole of nature accessible to us forms a system, an interconnected totality of bodies, and by bodies we understand here all material existence extending from stars to atoms, indeed right to ether particles, in so far as one grants the existence of the last named.   In the fact that these bodies are interconnected is already included that they react on one another, and it is precisely this mutual reaction that constitutes motion. It already becomes evident here that matter is unthinkable without motion.[2]   And if, in addition, matter confronts us as something given, equally uncreatable as indestructible, it follows that motion also is as uncreatable as indestructible. It became impossible to reject this conclusion as soon as it was recognised that the universe is a system, an interconnection of bodies. And since this recognition had been reached by philosophy long before it came into effective operation in natural science, it is explicable why philosophy, fully

[1] The nature of many chemical and electrical processes in the animal body is now well understood.

[2] Physicists who had not read Engels were startled by the recent discovery that even in the neighbourhood of the absolute zero of heat, atoms are still in vigorous internal motion.

two hundred years before natural science, drew the conclusion of the uncreatability and indestructibility of motion. Even the form in which it did so is still superior to the present day formulation of natural science. Descartes' principle, that the amount of motion present in the universe is always the same, has only the formal defect of applying a finite expression to an infinite magnitude. On the other hand, two expressions of the same law are at present current in natural science : Helmholtz's law of the conservation of *force*, and the newer, more precise, one of the conservation of *energy*. Of these, the one, as we shall see, says the exact opposite of the other, and moreover each of them expresses only one side of the relation.

When two bodies act on each other so that a change of place of one or both of them results, this change of place can consist only in an approach or a separation. They either attract each other or they repel each other. Or, as mechanics expresses it, the forces operating between them are central, acting along the line joining their centres. That this happens, that it is the case throughout the universe without exception, however complicated many movements may appear to be, is nowadays accepted as a matter of course. It would seem nonsensical to us to assume, when two bodies act on each other and their mutual interaction is not opposed by any obstacle or the influence of a third body, that this action should be effected otherwise than along the shortest and most direct path, *i.e.* along the straight line joining their centres. It is well known, moreover, that Helmholtz (*Erhaltung der Kraft* [*The Conservation of Force*], Berlin, 1847, Sections 1 and 2) has provided the mathematical proof that central action and unalterability of the quantity of motion are reciprocally conditioned and that the assumption of other than central actions leads to results in which motion could

be either created or destroyed. Hence the basic form
of all motion is approximation and separation, contraction and expansion—in short, the old polar opposites
of *attraction* and *repulsion*.

It is expressly to be noted that attraction and repulsion
are not regarded here as so-called "*forces*" but as
simple forms of motion, just as Kant had already conceived matter as the unity of attraction and repulsion.
What is to be understood by the conception of " forces "
will be shown in due course.

All motion consists in the interplay of attraction and
repulsion. Motion, however, is only possible when each
individual attraction is compensated by a corresponding
repulsion somewhere else. Otherwise in time one side
would get the preponderance over the other and then
motion would finally cease. Hence all attractions and
all repulsions in the universe must mutually balance one
another. Thus the law of the indestructibility and un-
creatibility of motion takes the form that each movement
of attraction in the universe must have as its complement
an equivalent movement of repulsion and *vice versa* ;
or, as ancient philosophy—long before the natural
scientific formulation of the law of conservation of force
or energy—expressed it : the sum of all attractions in
the universe is equal to the sum of all repulsions.

However it appears that there are still two possibilities
for all motion to cease at some time or other, either by
repulsion and attraction finally cancelling each other
out in actual fact, or by the total repulsion finally taking
possession of one part of matter and the total attraction
of the other part. For the dialectical conception, these
possibilities are excluded from the outset. Dialectics
has proved from the results of our experience of nature
so far that all polar opposites in general are determined
by the mutual action of the two opposite poles on
one another, that the separation and opposition of these

poles exists only within their unity and inter-connection, and, conversely, that their inter-connection exists only in their separation and their unity only in their opposition. This once established, there can be no question of a final cancelling out of repulsion and attraction, or of a final partition between the one form of motion in one half of matter and the other form in the other half, consequently there can be no question of mutual penetration or of absolute separation of the two poles. It would be equivalent to demanding in the first case that the north and south poles of a magnet should mutually cancel themselves out or, in the second case, that dividing a magnet in the middle between the two poles should produce on one side a north half without a south pole, and on the other side a south half without a north pole. Although, however, the impermissibility of such assumptions follows at once from the dialectical nature of polar opposites, nevertheless, thanks to the prevailing metaphysical mode of thought of natural scientists, the second assumption at least plays a certain part in physical theory. This will be dealt with in its place.

How does motion present itself in the interaction of attraction and repulsion? We can best investigate this in the separate forms of motion itself. At the end, the general aspect of the matter will show itself.

Let us take the motion of a planet about its central body. The ordinary school textbook of astronomy follows Newton in explaining the ellipse described as the result of the joint action of two forces, the attraction of the central body and a tangential force driving the planet along the normal to the direction of this attraction. Thus it assumes, besides the form of motion directed centrally, also another direction of motion or so-called " force " perpendicular to the line joining the central points. Thereby it contradicts the above-mentioned basic law according to which all motion in our universe can only take place along the line joining the central points of the bodies acting on one another, or, as one says, is caused only by centrally acting forces. Equally, it introduces into the theory an element of motion which, as we have likewise seen, necessarily leads to the creation and destruction of motion, and therefore presupposes a creator. What had to be done, therefore, was to reduce this mysterious tangential force to a form of motion acting centrally, and this the Kant-Laplace theory of cosmogony accomplished. As is well known, according to this conception the whole solar system arose from a rotating, extremely tenuous, gaseous mass by gradual contraction. The rotational motion is obviously strongest at the equator of this gaseous sphere, and individual gaseous rings separate themselves from the mass and clump themselves together into planets, planetoids, etc., which revolve round the central body in the direction of the original rotation. This rotation itself is usually explained from the motion characteristic of the individual particles of gas. This motion takes place in all directions, but finally an excess in one particular direction makes itself evident and so causes the rotating motion, which is bound to become stronger and stronger with the progressive contraction of the gaseous sphere. But whatever hypothesis is assumed of the origin of the rotation, it abolishes the tangential force, dissolving it in a special form of the phenomena of centrally acting motion. If the one element of planetary motion, the directly central one, is represented by gravitation, the attraction between the planet and the central body, then the other tangential element appears as a relic, in a derivative or altered form, of the original repulsion of the individual particles of the gaseous sphere. Then the life process of a solar system presents itself as an interplay of attraction and repulsion, in which attraction gradually more and more gets the upper hand owing to repulsion being radiated into space in the form of heat and thus more and more becoming lost to the system.

One sees at a glance that the form of motion here conceived as repulsion is the same as that which modern physics terms " *energy.*" By the contraction of the system and the resulting detachment of the individual bodies of which it consists to-day, the system has lost " energy," and indeed this loss, according to Helmholtz's well-known calculation,[1] already amounts to 453/454 of the total quantity of motion originally present in the form of repulsion.

Let us take now a mass in the shape of a body on our earth itself. It is connected with the earth by gravitation, as the earth in turn is with the sun ; but unlike the earth it is incapable of a free planetary motion. It can be set in motion only by an impulse from outside, and even then, as soon as the impulse ceases, its movement speedily comes to a standstill, whether by the effect of gravity alone or by the latter in combination with the resistance of the medium in which it moves. This resistance also is in the last resort an effect of gravity, in the absence of which the earth would not have on its surface any resistant medium, any atmosphere. Hence in pure mechanical motion on the earth's surface we are concerned with a situation in which gravitation, attraction, decisively predominates, where therefore the production of the motion shows both phases : first counteracting gravity and then allowing gravity to act— in a word, production of rising and falling.

Thus we have again mutual action between attraction on the one hand and a form of motion taking place in

---

[1] Since Helmholtz's time the huge attractive forces between certain atomic nuclei have been discovered. If these are taken into account the loss is far less.

the opposite direction to it, hence a repelling form of motion, on the other hand. But within the sphere of terrestrial *pure* mechanics (which deals with masses of *given* states of aggregation and cohesion taken by it as unalterable) this repelling form of motion does not occur in nature. The physical and chemical conditions under which a lump of rock becomes separated from a mountain top, or a fall of water becomes possible, lie outside our sphere. Therefore, in terrestrial pure mechanics, the repelling, raising motion must be produced artificially : by human force, animal force, water or steam power, etc. And this circumstance, this necessity to combat the natural attraction artificially, causes the mechanicians to adopt the view that attraction, gravitation, or, as they say, the force of gravity, is the most important, indeed the basic, form of motion in nature.

When, for instance, a weight is raised and communicates motion to other bodies by falling directly or indirectly, then according to the usual view of mechanics it is not the *raising* of the weight which communicates this motion but the *force of gravity*. Thus Helmholtz, for instance, makes " the force which is the simplest and the one with which we are best acquainted, viz. gravity, act as the driving force . . . for instance in grandfather clocks that are actuated by a weight. The weight . . . cannot comply with the pull of gravity without setting the whole clockwork in motion." But it cannot set the clockwork in motion without itself sinking and it goes on sinking until the string from which it hangs is completely unwound :

" Then the clock comes to a stop, for the operative capacity of the weight is exhausted for the time being. Its weight is not lost or diminished, it remains attracted to the same extent by the earth, but the capacity of this weight to produce movements has been lost. . . . We can, however, wind up the clock by the power of the human arm, whereby the weight is once more raised up. As soon as this has happened, it regains its previous operative capacity and can again keep the clock in motion." (Helmholtz, *Popular Lectures*, German Edition, II. pp. 144–5.)

According to Helmholtz, therefore, it is not the active communication of motion, the raising of the weight, that sets the clock into motion, but the passive heaviness of the weight, although this same heaviness is only withdrawn from its passivity by the raising, and once again returns to passivity after the string of the weight has unwound. If then according to the modern conception, as we saw above, *energy* is only another expression for *repulsion*, here in the older Helmholtz conception *force* appears as another expression for the opposite of repulsion, for *attraction*. For the time being we shall simply put this on record.

When this process, as far as terrestrial mechanics is concerned, has reached its end, when the heavy mass has first of all been raised and then again let fall through the same height, what becomes of the motion that constituted it ? For pure mechanics, it has disappeared. But we know now that it has by no means been destroyed. To a lesser extent it has been converted into the air oscillations of sound waves, to a much greater extent into heat—which has been communicated in part to the resisting atmosphere, in part to the falling body itself, and finally in part to the floor, on which the weight comes to rest. The clock weight has also gradually given up its motion in the form of frictional heat to the separate driving wheels of the clockwork. But, although usually expressed in this way, it is not the *falling* motion, *i.e.* the attraction, that has passed into heat, and therefore into a form of repulsion. On the contrary, as Helmholtz correctly remarks, the attraction, the heaviness, remains what it previously was and, accurately speaking, becomes even greater. Rather it is the repulsion communicated to the raised body by raising that is *mechanically* destroyed by falling and reappears as heat. The repulsion of masses is transformed into molecular repulsion.

Heat, as already stated, is a form of repulsion. It sets the molecules of solid bodies into oscillation, thereby loosening the connections of the separate molecules until finally the transition to the liquid state takes place. In the liquid state also, on continued addition of heat, it increases the motion of the molecules until a degree is reached at which the latter split off altogether from the mass and, at a definite velocity determined for each molecule by its chemical constitution, they move away individually in the free state. With a still further addition of heat, this velocity is further increased, and so the molecules are more and more repelled from one another.

But heat is a form of so-called " energy " ; here once again the latter proves to be identical with repulsion.

In the phenomena of static electricity and magnetism, we have a polar division of attraction and repulsion. Whatever hypothesis may be adopted of the *modus operandi* of these two forms of motion, in view of the facts no one has any doubt that attraction and repulsion, in so far as they are produced by static electricity or magnetism and are able to develop unhindered, completely compensate one another, as in fact necessarily follows from the very nature of the polar division. Two poles whose activities did not completely compensate each other would indeed not be poles, and also have so far not been discovered in nature. For the time being we will leave galvanism out of account, because in its case the process is determined by chemical reactions, which makes it more complicated. Therefore, let us investigate rather the chemical processes of motion themselves.

When two parts by weight of hydrogen combine with 15·96 parts by weight of oxygen to form water vapour, an amount of heat of 68,924 heat units is developed during the process. Conversely, if 17·96 parts by weight of water vapour are to be decomposed into 2 parts by weight of hydrogen and 15·96 parts by weight of oxygen, this is only possible on condition that the water vapour has communicated to it an amount of motion equivalent to 68,924 heat units—whether in the form of heat itself or of electrical motion. The same thing holds for all other chemical processes. In the overwhelming majority of cases, motion is given off on combination and must be supplied on decomposition. Here, too, as a rule, repulsion is the active side of the process more endowed with motion or requiring the addition of motion, while attraction is the passive side producing a surplus of motion and giving off motion. On this account, the modern theory also declares that, on the whole, energy is set free on the combination of elements and is bound up on decomposition. And Helmholtz declares :

" This force (chemical affinity) can be conceived as a force of *attraction*. . . . This force of attraction between the atoms of carbon and oxygen performs work quite as much as that exerted on a raised weight by the earth in the form of gravitation. . . . When carbon and oxygen atoms rush at one another and combine to form carbonic acid,[1] the newly-formed particles of carbonic acid must be in very violent molecular motion, *i.e.* in heat motion. . . . When later they have given up their heat to the environment, we still have in the carbonic acid all the carbon,

[1] Now usually called carbon dioxide.

all the oxygen, and in addition the affinity of both continuing to exist just as powerfully as before. But this affinity now expresses itself solely in the fact that the atoms of carbon and oxygen stick fast to one another, and do not allow of their being separated " (Helmholtz, *loc. cit.*, p. 169).

It is just as before : Helmholtz insists that in chemistry as in mechanics *force* consists only in *attraction*, and therefore is the exact opposite of what other physicists call energy and which is identical with *repulsion*.

Hence we have now no longer the two simple basic forms of attraction and repulsion, but a whole series of sub-forms in which the winding up and running down process of universal motion goes on in opposition to both attraction and repulsion. It is, however, by no means merely in our mind that these manifold forms of appearance are comprehended under the single expression of motion. On the contrary, they themselves prove in action that they are forms of one and the same motion by passing into one another under given conditions. Mechanical motion of masses passes into heat, into electricity, into magnetism ; heat and electricity pass into chemical decomposition ; chemical combination in turn develops heat and electricity and, by means of the latter, magnetism ; and finally, heat and electricity produce once more mechanical movement of masses. Moreover, these changes take place in such a way that a given quantity of motion of one form always has corresponding to it an exactly fixed quantity of another form. Further, it is a matter of indifference which form of motion provides the unit by which the amount of motion is measured, whether it serves for measuring mass motion, heat, so-called electromotive force, or the motion undergoing transformation in chemical processes.

We base ourselves here on the theory of the " conservation of energy " established by J. R. Mayer [1] in 1842 and afterwards worked out internationally with such brilliant success, and we have now to investigate the fundamental concepts nowadays made use of by this theory. These are the concepts of " force," " energy," and " work."

It has been shown above that according to the modern view, now fairly generally accepted, energy is the term used for repulsion, while Helmholtz generally uses the word force to express attraction. One could regard this as a mere distinction of form, inasmuch as attraction and repulsion compensate each other in the universe, and accordingly it would appear a matter of indifference which side of the relation is taken as positive and which as negative, just as it is of no importance in itself whether the positive abscissæ are counted to the right or the left of a point in a given line. Nevertheless, this is not absolutely so.

For we are concerned here, first of all, not with the universe, but with phenomena occurring on the earth

[1] Helmholtz, in his *Pop. Vorlesungen* [*Popular Lectures*], II, p. 113, appears to ascribe a certain share in the natural scientific proof of Descartes' principle of the quantitative immutability of motion to himself as well as to Mayer, Joule, and Colding. " I myself, without knowing anything of Mayer and Colding, and only becoming acquainted with Joule's experiments at the end of my work, *proceeded along the same path* ; I occupied myself especially with searching out all the relations between the various processes of nature that could be deduced from the given mode of consideration, and I *published my investigations in* 1847 in a little work entitled *Uber die Erhaltung der Kraft* [*On the Conservation of Force*]."—But in this work there is to be found nothing new for the position in 1847 beyond the above-mentioned, mathematically very valuable, development that " conservation of force " and central action of the forces active between the various bodies of a system are only two different expressions for the same thing, and further a more accurate formulation of the law that the sum of the live and tensional forces * in a given *mechanical* system is constant. In every other respect it was already superseded since Mayer's second paper of 1845. Already in 1842 Mayer maintained the " indestructibility of force," and from his new standpoint in 1845 he had much more brilliant things to say about the " relations between the various processes of nature " than Helmholtz had in 1847. [*Note by F. Engels.*]

* " Live force " or *vis viva* is now termed kinetic energy, and " tensional force " potential energy.

and conditioned by the exact position of the earth in the solar system, and of the solar system in the universe. At every moment, however, our solar system gives out enormous quantities of motion into space, and motion of a very definite quality, viz. the sun's heat, *i.e.* re- pulsion.[1] But our earth itself allows of the existence of life on it only owing to the sun's heat, and it in turn finally radiates into space the sun's heat received, after it has converted a portion of this heat into other forms of motion. Consequently, in the solar system and above all on the earth, attraction already considerably pre- ponderates over repulsion. Without the repulsive motion radiated to us from the sun, all motion on the earth would cease. If to-morrow the sun were to become cold, the attraction on the earth would still, other circumstances remaining the same, be what it is to-day. As before, a stone of 100 kilogrammes, wherever situated, would weigh 100 kilogrammes. But the motion, both of masses and of molecules and atoms, would come to what we would regard as an absolute standstill. Therefore it is clear that for processes occurring on the *earth* to-day it is by no means a matter of indifference whether attraction or repulsion is con- ceived as the active side of motion, hence as " force " or " energy." On the contrary, on the earth to-day attraction has already become *altogether passive* owing to its decisive preponderance over repulsion ; we owe all active motion to the supply of repulsion from the sun. Therefore, the modern school—even if it remains unclear about the nature of the relation constituting motion—nevertheless, in point of fact and for *terrestrial* processes, indeed for the whole solar system, is abso- lutely right in conceiving energy as repulsion.

[1] Again Engels was ahead of his time. It was only in 1900 that radiant heat and light were shown by Lebedeff to exercise repulsion on the bodies emitting, absorbing, or reflecting them.

The expression " energy " by no means correctly expresses all the relationships of motion, for it comprehends only one aspect, the action but not the reaction. It still makes it appear as if " energy " was something external to matter, something implanted in it. But in all circumstances it is to be preferred to the expression " force."

As conceded on all hands (from Hegel to Helmholtz), the notion of force is derived from the activity of the human organism within its environment. We speak of muscular force, of the lifting force of the arm, of the leaping power of the legs, of the digestive force of the stomach and intestinal canal, of the sensory force of the nerves, of the secretory force of the glands, etc. In other words, in order to save having to give the real cause of a change brought about by a function of our organism, we fabricate a fictitious cause, a so-called force corresponding to the change. Then we carry this convenient method over to the external world also, and so invent as many forces as there are diverse phenomena.[1]

In *Hegel's* time natural science (with the exception perhaps of heavenly and terrestrial mechanics) was still in this naïve state, and Hegel quite correctly attacks the prevailing way of denoting forces (passage to be quoted).[2] Similarly in another passage :

" It is better (to say) that a magnet has a *Soul* (as Thales expresses it) than that it has an attracting force ; force is a kind of property which is *separable from matter* and put forward as a predicate—while soul, on the other hand, *is its movement, identical with the nature of matter.*" (*Geschichte der Philosophie* [*History of Philosophy*], I, p. 208.)

To-day we no longer make it so easy for ourselves in regard to forces. Let us listen to Helmholtz :

[1] Nowadays this tendency has been reversed. No one but an extreme vitalist would speak of a secretory force in a gland. The saliva, for example, appears to be separated from the blood by forces which are essentially electrical. [2] See Appendix II, p. 331.

" If we are fully acquainted with a natural law, we must also demand that it should operate without exception. . . . Thus the law confronts us as an objective power, and accordingly we term it a *force*. For instance, we objectivise the law of the refraction of light as a refractive power of transparent substances, the law of chemical affinities as a force of affinity of the various substances for one another. Thus we speak of the electrical force of contact of metals, of the force of adhesion, capillary force, and so on. These names objectivise laws which in the first place embrace only a limited series of natural processes, *the conditions for which are still rather complicated.* . . . Force is only the objectivised law of action. . . . The abstract idea of force introduced by us only makes the addition that we have not arbitrarily invented this law but that it is a compulsory law of phenomena. Hence our demand to *understand* the phenomena of nature, *i.e.* to find out their *laws*, takes on another form of expression, viz. that we have to seek out the *forces* which are the causes of the phenomena." (*Loc. cit.*, pp. 189–191. Innsbruck lecture of 1869.)

Firstly, it is certainly a peculiar manner of " objectivising" if the *purely subjective* notion of *force* is introduced into a natural law that has already been established as independent of our subjectivity and therefore completely *objective*. At most an Old-Hegelian of the strictest type might permit himself such a thing, but not a Neo-Kantian like Helmholtz. Neither the law, when once established, nor its objectivity, nor that of its action, acquires the slightest new objectivity by our interpolating a force into it ; what is added is our *subjective assertion* that it acts in virtue of some so far entirely unknown force. The secret meaning, however, of this interpolating is seen as soon as Helmholtz gives us examples : refraction of light, chemical affinity, contact electricity, adhesion, capillarity, and confers

on the laws that govern these phenomena the " objective"
honorary rank of *forces*. " These names objectivise
laws which in the first place embrace only a limited
series of natural processes, the conditions for which
are still rather complicated." And it is just here that the
" objectivising," which is rather subjectivising, gets its
meaning ; not because we have become fully acquainted
with the law, but just because this is *not* the case. Just
because we are *not* yet clear about the " rather complicated conditions " of these phenomena, we often resort
here to the word force. We express thereby not our
scientific knowledge, but our *lack* of scientific knowledge
of the nature of the law and its mode of action. In this
sense, as a short expression for a causal connection that
has not yet been explained, as a makeshift expression, it
may pass for current usage. Anything more than that
is bad. With just as much right as Helmholtz explains
physical phenomena from so-called refractive force,
electrical force of contact, etc., the mediæval scholastics
explained temperature changes by means of a *vis calori-
fica* and a *vis frigifaciens* and thus saved themselves all
further investigation of heat phenomena.

And even in this sense it is one-sided, for it expresses
everything in a one-sided manner. All natural processes
are two-sided, they rest on the relation of at least two
effective parts, action and reaction. The notion of
force, however, owing to its origin from the action of
the human organism on the external world, and further
because of terrestrial mechanics, implies that only one
part is active, effective, the other part being passive,
receptive ; hence it lays down a not yet demonstrable
extension of the difference between the sexes to non-
living objects. The reaction of the second part, on
which the force works, appears at most as a passive
reaction, as a *resistance*. This mode of conception is
permissible in a number of fields even outside pure

mechanics, namely where it is a matter of the simple
transference of motion and its quantitative calculation.
But already in the more complicated physical processes
it is no longer adequate, as Helmholtz's own examples
prove. The refractive force lies just as much in the
light itself as in the transparent bodies. In the case of
adhesion and capillarity, it is certain that the " force "
is just as much situated in the surface of the solid as in
the liquid. In contact electricity, at any rate, it is
certain that *both* metals contribute to it, and " chemical
affinity " also is situated, if anywhere, in *both* the parts
entering into combination. But a force which consists
of separated forces, an action which does not evoke
its reaction, but which exists solely by itself, is no force
in the sense of terrestrial mechanics, the only science in
which one really knows what is meant by a force.
For the basic conditions of terrestrial mechanics are,
firstly, refusal to investigate the causes of the impulse,
i.e. the nature of the particular force, and, secondly,
the view of the one-sidedness of the force, it being
everywhere opposed by an identical gravitational
force, such that in comparison with any terrestrial
distance of fall the earth's radius $= \infty$.

But let us see further how Helmholtz " objectivises "
his " forces " into natural laws.

In a lecture of 1854 (*loc. cit.*, p. 119) he examines the
" store of working force " [1] originally contained in the
nebular sphere from which our solar system was formed.
" In point of fact it received an enormously large legacy
in this respect, if only in the form of the general force
of attraction of all its parts for one another." This
indubitably is so. But it is equally indubitable that the
whole of this legacy of gravitation is present undiminished
in the solar system to-day, apart perhaps from the
minute quantity that was lost together with the matter

[1] We should now call this potential energy.

which was flung out, possibly irrevocably, into space. Further, " The chemical forces too must have been already present and ready to act ; but as these forces could become effective only on intimate contact of the various kinds of masses, condensation had to take place before they came into play." If, as Helmholtz does above, we regard these chemical forces as forces of affinity, hence as *attraction*, then again we are bound to say that the sum-total of these chemical forces of attraction still exists undiminished within the solar system.

But on the same page Helmholtz gives us the results of his calculations " that perhaps only the 454th part of the original mechanical force exists as such "—that is to say, in the solar system. How is one to make sense of that ? The force of attraction, general as well as chemical, is still present unimpaired in the solar system. Helmholtz does not mention any other certain source of force. In any case, according to Helmholtz, these forces have performed tremendous work. But they have neither increased nor diminished on that account. As it is with the clock weight mentioned above, so it is with every molecule in the solar system and with the solar system itself. " Its gravitation is neither lost nor diminished." What happens to carbon and oxygen as previously mentioned holds good for all chemical elements : the total given quantity of each one remains, and " the total force of affinity continues to exist just as powerfully as before." What have we lost then ? And what " force " has performed the tre- mendous work which is 453 times as big as that which, according to his calculation, the solar system is still able to perform ? Up to this point Helmholtz has given no answer. But further on he says :

" Whether a further *reserve of force in the shape of heat* was present, we do not know."—But, if we may be allowed to mention it, heat is a repulsive " force," it

acts therefore *against* the direction of both gravitation
and chemical attraction, being minus if these are put as
plus. Hence if, according to Helmholtz, the original
store of force is composed of general and chemical
attraction, an extra reserve of heat would have to be,
not added to that reserve of force, but subtracted from
it. Otherwise the sun's heat would have had to *strengthen*
the force of attraction of the earth when it causes water
to evaporate in direct opposition to this attraction, and
the water vapour to rise ; or the heat of an incandescent
iron tube through which steam is passed would *strengthen*
the chemical attraction of oxygen and water, whereas
it puts it out of action. Or, to make the same thing
clear in another form : let us assume that the nebular
sphere with radius r, and therefore with volume $\frac{4}{3}\pi r^3$,
has a temperature t. Let us further assume a second
nebular sphere of equal mass having at the higher
temperature T the larger radius R and volume $\frac{4}{3}\pi R^3$.
Now it is obvious that in the second nebular sphere the
attraction, mechanical as well as physical and chemical,
can act with the same force as in the first only when it
has shrunk from radius R to radius r, *i.e.* when it has
radiated into world space heat corresponding to the
temperature difference T—t. A hotter nebular sphere
will therefore condense later than a colder one ; consequently the heat, considered from Helmholtz's standpoint
as an obstacle to condensation, is no plus but a minus
of the " reserve of force." Helmholtz, by pre-supposing
the possibility of a quantum of *repulsive* motion in the
form of heat becoming added to the *attractive* forms of
motion and increasing the total of these latter, commits
a definite error of calculation.

Let us now bring the whole of this " reserve of force,"
possible as well as demonstrable, under the same
mathematical sign so that an addition is possible. Since
for the time being we cannot reverse the heat and re-

place its repulsion by the equivalent attraction, we shall
have to perform this reversal with the two forms of
attraction. Then, instead of the general force of
attraction, instead of the chemical affinity, and instead
of the heat, which moreover possibly already exists as
such at the outset, we have simply to put—the sum
of the repulsive motion or so-called energy present in the
gaseous sphere at the moment when it becomes independent. And by so doing Helmholtz's calculation will
also hold, in which he wants to calculate " the heating
that must arise from the assumed initial condensation
of the heavenly bodies of our system from nebulously
scattered matter." By thus reducing the whole " reserve
of force " to heat, repulsion, he also makes it possible
to add on the assumed " heat reserve force." The
calculation then asserts that 453/454 of all the energy,
i.e. repulsion, originally present in the gaseous sphere has
been radiated into space in the form of heat, or, to put it
accurately, that the sum of all attraction in the present
solar system is to the sum of all repulsion, still present
in the same, as 453 : 1. But then it directly contradicts the text of the lecture to which it is added as proof.

If then the notion of force, even in the case of a
physicist like Helmholtz, gives rise to such confusion of
ideas, this is the best proof that it is in general not
susceptible of scientific use in all branches of investigation which go beyond the calculations of mechanics.
In mechanics the causes of motion are taken as given
and their origin is disregarded, only their effects being
taken into account. Hence if a cause of motion is
termed a force, this does no damage to mechanics as
such ; but it becomes the custom to transfer this term
also to physics, chemistry, and biology, and then
confusion is inevitable. We have already seen this and
shall frequently see it again.

For the concept of work, see the next chapter.

IV

THE MEASURE OF MOTION—WORK [1]

" On the other hand, I have always found hitherto that the basic concepts in this field (*i.e.* " the basic physical concepts of work and their unalterability ") seem very difficult to grasp for persons who have not gone through the school of mathematical mechanics, in spite of all zeal, all intelligence, and even a fairly high degree of scientific knowledge. Moreover, it cannot be denied that they are abstractions of a quite peculiar kind. It was not without difficulty that even such an intellect as that of I. Kant succeeded in understanding them, as is proved by his polemic against Leibniz on this subject."

So says Helmholtz (*Pop. wiss. Vorträge* [*Popular Scientific Lectures*], II, Preface).

According to this, we are venturing now into a very dangerous field, the more so since we cannot very well take the liberty of guiding the reader " through the school of mathematical mechanics." Perhaps, however, it will turn out that, where it is a question of concepts, dialectical thinking will carry us at least as far as mathematical calculation.

Galileo discovered, on the one hand, the law of falling, according to which the distances traversed by falling bodies are proportional to the squares of the times taken in falling. On the other hand, as we shall see, he put

---

[1] In the physics of the last fifty years the notion of force has become rather unimportant as compared with that of energy. Hence a good deal of this chapter, while important at the time when it was written, is nowadays less so. This does not, of course, diminish its interest as a penetrating criticism of nineteenth-century physics. Indeed, we shall see that Engels pointed out some of the lines along which physics actually advanced.

forward the not quite compatible law that the magnitude of motion of a body (its *impeto* or *momento*) is determined by the mass and the velocity in such a way that for constant mass it is proportional to the velocity. Descartes adopted this latter law and made the product of the mass and the velocity of the moving body quite generally into the measure of its motion.

Huyghens had already found that, on elastic impact, the sum of the products of the masses, multiplied by the squares of their velocities, remains the same before and after impact, and that an analogous law holds good in various other cases of motion to a system of connected bodies.

Leibniz was the first to realise that the Cartesian measure of motion was in contradiction to the law of falling. On the other hand, it could not be denied that in many cases the Cartesian measure was correct. Accordingly, Leibniz divided moving forces into dead forces and live forces.[1] The dead were the " pushes "

[1] In a system of moving bodies certain quantities remain constant provided the system is not acted on from outside and does not act on outside bodies. (Of course no real system is completely isolated.) These quantities include its mass, and seven others which depend on the motions of the bodies. One of these is the energy of the system. It consists of two parts, namely that due to the actual motion of bodies whether visible (ordinary motion) or too small to be seen (heat), and that due to the fact that the bodies may acquire more motion if their mutual repulsion or attraction is allowed to do work on them. These are called kinetic and potential energy. For example, a falling bomb has kinetic energy depending on its actual speed, and potential energy due both to its height above the earth, which enables it to gain more speed, and to the fact that when it explodes the atoms in its charge of explosives will re-arrange themselves, and thus put its parts into violent motion. The kinetic energy of a moving body is proportional to its mass and the square of its velocity.

Besides this the amount of momentum in the system remains constant. Momentum is proportional to mass multiplied by velocity, and has a direction, which must be specified before it is measured. Thus if a one-ton lorry moving south at 20 m.p.h. has +20 units of momentum, a similar lorry moving north at the same speed has −20 units, whereas energy is essentially positive, and two energies cannot cancel out. If the lorries collide they come to rest, and the total momentum is zero as before. But their energy is mostly converted into heat and sound. Momentum must be measured in three directions at right-angles, or "pulls" of resting bodies, and their measure the product of the mass and the velocity with which the body would move if it were to pass from a state of rest to one of motion. On the other hand, he put forward as the measure of *vis viva*, of the real motion of a body, the product of the mass and the square of the velocity. This new measure of motion he derived directly from the law of falling.

"The same force is required," so Leibniz concluded, "to raise a body of four pounds in weight one foot as to raise a body of one pound in weight four feet; but the distances are proportional to the square of the velocity, for when a body has fallen four feet, it attains twice the velocity reached on falling only one foot. However, bodies on falling acquire the force for rising to the same height as that from which they fell; hence the forces are proportional to the square of the velocity." (Suter, *Geschichte der Mathematik* [*History of Mathematics*], II, p. 367.)

But he showed further that the measure of motion *mv* is in contradiction to the Cartesian law of the constancy of the quantity of motion, for if it was really

e.g. south, east, and up. If so, momentum in each direction is conserved, that is to say remains constant in the system as a whole.

Besides this, moment of momentum, also called angular momentum, which may be regarded as a measure of spin, is conserved. This again must be measured about axes in three directions at right-angles to one another. Engels is not concerned with angular momentum in this chapter.

Now Descartes recognised the law of the conservation of momentum in certain cases, while Leibniz saw that in the absence of friction the sum of potential and kinetic energies remained constant.

Leibniz's measure of kinetic energy was mv^2, a quantity which he called the "lebendige Kraft" or "live force." This was called "*vis viva*" by English writers, and will be so translated here. It has, of course, nothing to do with the so-called "vital force" in living creatures, which has never been observed, much less measured.

Of the seven quantities which remain constant in a moving system, only energy can be regarded as a measure of motion, not only for the reasons given by Engels, but because the other six have directions, and indeed are only known when the directions are known. Whereas a measure is directionless, as for example a foot northwards is equal to a foot southwards.

valid the force (*i.e.* the quantity of motion) in nature would continually increase or diminish. He even devised an apparatus (1690, *Acta Eruditorum*) which, if the measure *mv* were correct, would be bound to act as a *perpetuum mobile* with continual gain of force, which, however, would be absurd. Recently, Helmholtz has again frequently employed this kind of argument.

The Cartesians protested with might and main and there developed a famous controversy lasting many years, in which Kant also participated in his very first work (*Gedanken von der wahren Schätzung der lebendigen Kräfte* [*Thoughts on the True Estimation of Live Forces*], 1746), without, however, seeing clearly into the matter. Mathematicians to-day look down with a certain amount of scorn on this " barren " controversy which " dragged out for more than forty years and divided the mathematicians of Europe into two hostile camps, until at last d'Alembert by his *Traité de dynamique* (1743), as it were by a final verdict, put an end to the *useless verbal dispute*, for it was nothing else." (Suter, *ibid.*, p. 366.)

It would, however, seem that a controversy could not rest entirely on a useless verbal dispute when it had been initiated by a Leibniz against a Descartes, and had occupied a man like Kant to such an extent that he devoted to it his first work, a fairly large volume. And in point of fact, how is it to be understood that motion has two contradictory measures, that on one occasion it is proportional to the velocity, and on another to the square of the velocity ? Suter makes it very easy for himself ; he says both sides were right and both were wrong ; " nevertheless, the expression ' *vis viva* ' has endured up to the present day ; *only it no longer serves as the measure of force*, but is merely a term that was once adopted for the product of the mass and half the square of the velocity, a product so full of significance in mechanics." Hence, *mv* remains the measure of motion, and *vis viva* is only another expression for $\dfrac{mv^2}{2}$, concerning which formula we learn indeed that it is of great significance for mechanics, but now most certainly do not know what significance it has.

Let us, however, take up the salvation-bringing *Traité de dynamique* and look more closely at d'Alembert's " final verdict " ; it is to be found in the *preface*. In the text, it says, the whole question does not occur, on account of *l'inutilité parfaite dont elle est pour la mécanique*.[1] This is quite correct for *purely mathematical* mechanics, in which, as in the case of Suter above, words used as designations are only other expressions, or names, for algebraic formulæ, names in connection with which it is best not to think at all. Nevertheless, since such important people have concerned themselves with the matter, he desires to examine it briefly in the preface. Clearness of thought demands that by the force of moving bodies one should understand only their property of overcoming obstacles or resisting them. Hence, force is to be measured neither by mv nor by mv^2, but solely by the obstacles and the resistance they offer.

Now, there are, he says, three kinds of obstacles : (1) insuperable obstacles which totally destroy the motion, and for that very reason cannot be taken into account here ; (2) obstacles whose resistance suffices to arrest the motion and to do so instantaneously : the case of equilibrium ; (3) obstacles which only gradually arrest the motion : the case of retarded motion.

" Or tout le monde convient qu'il y a équilibre entre deux corps, quand les produits de leurs masses par leurs vitesses virtuelles, c'est à dire par les vitesses avec lesquelles ils tendent à se mouvoir, sont égaux de part et d'autre. Donc dans l'équilibre le produit

[1] Its absolute uselessness for mechanics.

de la masse par la vitesse, ou, ce qui est la même chose,
la quantité de mouvement, peut représenter la force.
Tout le monde convient aussi que dans le mouvement
retardé, le nombre des obstacles vaincus est comme le
carré de la vitesse, en sorte qu'un corps qui a fermé
un ressort, par exemple, avec une certaine vitesse,
pourra, avec une vitesse double, fermer ou tout à la
fois, ou successivement, non pas deux, mais quatre
ressorts semblables au premier, neuf avec une vitesse
triple, et ainsi du reste.   D'où les partisans des
forces vives [the Leibnizians] concluent que la force
des corps qui se meuvent actuellement, est en général
comme le produit de la masse par le carré de la vitesse.
Au fond, quel inconvénient pourrait-il y avoir, à ce
que la mesure des forces fût différente dans l'équilibre
et dans le mouvement retardé, puisque, si on veut ne
raisonner que d'après des idées claires, on doit n'enten-
dre par le mot *force* que l'effet produit en surmontant
l'obstacle ou en lui résistant ? ''   (Preface, pp. 19–20,
of the original edition.) [1]

D'Alembert, however, is far too much of a philosopher
not to realise that the contradiction of a twofold measure
of one and the same force is not to be got over so easily.
Therefore, after repeating what is basically only the
same thing as Leibniz had already said—for his
équilibre is precisely the same thing as the '' dead

[1] '' Everyone will agree that two bodies are in equilibrium when the
products of their masses and virtual velocities, that is to say the velocities with which they tend to move, are equal for each body.   Hence, in
equilibrium the product of the mass and the velocity, or, what is the
same thing, the quantity of motion, can represent the force.   Everyone will agree also that in retarded motion the number of obstacles
overcome is as the square of the velocity, such that, for instance, a
body which has compressed a spring, with a certain velocity, could,
with twice the velocity, compress simultaneously or successively not
two, but four, springs similar to the first, or nine with triple the velocity,
and so on.   Whence the partisans of *vis viva* (the Leibnizians) conclude
that the force of bodies actually in motion is in general the product of
the mass and the square of the velocity.   Basically, what incon-
venience could there be in forces being measured differently in equilibrium and in retarded motion since, if one wants to use only clear views
in reasoning, one should understand by the word *force* only the effect
produced in surmounting the obstacle or resisting it ? ''

pressure " of Leibniz—he suddenly goes over to the side
of the Cartesians and finds the following expedient :
the product *mv* can serve as a measure of force, even
in the case of delayed motion,

> " si dans ce dernier cas on mesure la force, non par la
> quantité absolue des obstacles, mais par la somme des
> résistances de ces mêmes obstacles. Car on ne
> saurait douter que cette somme des résistances ne
> soit proportionelle à la quantité du mouvement *mv*,
> puisque, de l'aveu de tout le monde, la quantité du
> mouvement que le corps perd à chaque instant, est
> proportionelle au produit de la résistance par la durée
> infiniment petite de l'instant, et que la somme de ces
> produits est evidemment la résistance totale."[1]

This latter mode of calculation seems to him the more
natural one, " car un obstacle n'est tel qu'en tant qu'il
résiste et c'est, à proprement parler, la somme des
résistances qui est l'obstacle vaincu ; d'ailleurs, en
estimant ainsi la force, on a l'avantage d'avoir pour
l'équilibre et pour le mouvement retardé une mesure
commune." [2]   Still, everyone can take that as he likes.
And so, believing he has solved the question, by
what, as Suter himself acknowledges, is a mathematical
blunder, he concludes with unkind remarks on the
confusion reigning among his predecessors, and asserts
that after the above remarks there is possible only a

---

[1] " If in this last case the force is measured, not by the absolute
quantity of obstacles, but by the sum of the resistances of these same
obstacles.   For it could not be doubted that this sum of the resistances
would be proportional to the quantity of motion (*mv*), since, by general
agreement, the quantity of motion lost by the body at each instant is
proportional to the product of the resistance and the infinitely small
duration of the instant, and the sum of these products evidently makes
up the total resistance."

[2] " For an obstacle is only such in as much as it offers resistance, and,
properly speaking, it is the sum of the resistances that constitutes the
obstacle overcome ; moreover, in estimating the force in this way, one
has the advantage of having a common measure for the equilibrium and
for the retarded motion."

very futile metaphysical discussion or a still more discreditable purely verbal dispute.

D'Alembert's proposal for reaching a reconciliation amounts to the following calculation :

A mass 1, with velocity 1, compresses 1 spring in unit time.

A mass 1, with velocity 2, compresses 4 springs, but requires two units of time ; *i.e.* only 2 springs per unit of time.

A mass 1, with velocity 3, compresses 9 springs in three units of time, *i.e.* only 3 springs per unit of time.

Hence if we divide the effect by the time required for it, we again come from mv^2 to mv.

This is the same argument that Catelan in particular had already employed against Leibniz ; it is true that a body with velocity 2 rises against gravity four times as high as one with velocity 1, but it requires double the time for it ; consequently the quantity of motion must be divided by the time, and $=2$, not $=4$. Curiously enough, this is also Suter's view, who indeed deprived the expression " *vis viva* " of all logical meaning and left it only a mathematical one. But this is natural. For Suter it is a question of saving the formula mv in its significance as sole measure of the quantity of motion ; hence logically mv^2 is sacrificed in order to arise again transfigured in the heaven of mathematics.

However, this much is correct : Catelan's argument provides one of the bridges connecting mv with mv^2, and so is of importance.

The mechanicians subsequent to d'Alembert by no means accepted his verdict, for his final verdict was indeed in favour of mv as the measure of motion. They adhered to his expression of the distinction which Leibniz had already made between dead and live forces : mv is valid for equilibrium, *i.e.* for statics ; mv^2 is valid for motion against resistance, *i.e.* for dynamics. Although on the whole correct, the distinction in this form has, however, logically no more meaning than the famous pronouncement of the junior officer : on duty always " to me," off duty always " me." It is accepted tacitly, it just exists. We cannot alter it, and if a contradiction lurks in this double measure, how can we help it ?

Thus, for instance, Thomson and Tait say (*A Treatise on Natural Philosophy*, Oxford, 1867, p. 162) ; " The *quantity of motion* or the *momentum* of a rigid body moving without rotation is proportional to its mass and velocity conjointly. Double mass or double velocity would correspond to double quantity of motion." And immediately below that they say : " The *vis viva* or *kinetic energy* of a moving body is proportional to the mass and the square of the velocity conjointly." [1]

The two contradictory measures of motion are put side by side in this very glaring form. Not so much as the slightest attempt is made to explain the contradiction, or even to disguise it. In the book by these two Scotsmen, thinking is forbidden, only calculation is permitted. No wonder that at least one of them, Tait, is accounted one of the most pious Christians of pious Scotland.

In Kirchhoff's *Vorlesungen über mathematische Mechanik* [*Lectures on Mathematical Mechanics*] the formulæ mv and mv^2 do not occur at all *in this form*.

Perhaps Helmholtz will aid us. In his *Erhaltung der Kraft* [*Conservation of Force*] he proposes to express *vis viva* by $\dfrac{mv^2}{2}$, a point to which we shall return later.

Then, on page 20 *et seq.*, he enumerates briefly the cases in which so far the principle of the conservation of

[1] See Appendix II, p. 332.

vis viva $\left(\text{hence of } \dfrac{mv^2}{2}\right)$ has been recognised and made use of.   Included therein under No. 2 is

> " the transference of motion by incompressible
> solid and fluid bodies, in so far as friction or impact
> of inelastic materials does not occur. For these
> cases our general principle is usually expressed in the
> rule that motion propagated and altered by mechanical
> powers always decreases in intensity of force in the
> same proportion as it increases in velocity. If,
> therefore, we imagine a weight *m* being raised with
> velocity *c* by a machine in which a force for per-
> forming work is produced uniformly by some process
> or other, then with a different mechanical arrange-
> ment the weight *nm* could be raised, but only with
> velocity *c/n*, so that in both cases the quantity of
> tensile force produced by the machine in unit time
> is represented by *mgc*, where *g* is the intensity of the
> gravitational force."

Thus, here too we have the contradiction that an
" intensity of force," which decreases and increases in
simple proportion to the velocity, has to serve as proof
for the conservation of an intensity of force which
decreases and increases in proportion to the square of the
velocity.

In any case, it becomes evident here that mv and mv^2
serve to determine two quite distinct processes, but we
certainly knew long ago that mv^2 cannot equal mv,
unless $v=1$.   What has to be done is to make it comprehensible why motion should have a twofold measure,
a thing which is surely just as unpermissible in natural
science as in commerce.   Let us, therefore, attempt this
in another way.

By mv, then, one measures " a motion propagated
and altered by mechanical powers " ; hence this measure
holds good for the lever and all its derivatives, for
wheels, screws, etc., in short, for all machinery for the

transference of motion. But from a very simple and by no means new consideration it becomes evident that in so far as mv applies here, so also does mv^2. Let us take any mechanical contrivance in which the sums of the lever-arms on the two sides are related to each other as 4 : 1, in which, therefore, a weight of 1 kg. holds a weight of 4 kg. in equilibrium. Hence, by a quite insignificant additional force on one arm of the lever we can raise 1 kg. by 20 m. ; the same additional force, when applied to the other arm of the lever, raises 4 kg. a distance of 5 m., and the preponderating weight sinks in the same time that the other weight requires for rising. Mass and velocity are inversely proportional to one another ; mv, $1 \times 20 = m'v'$, 4×5. On the other hand, if we let each of the weights, after it has been raised, fall freely to the original level, then the one, 1 kg., after falling a distance of 20 m. (the acceleration due to gravity is put in round figures $= 10$ m. instead of 9,81 m.), attains a velocity of 20 m. : the other, 4 kg., after falling a distance of 5 m., attains a velocity of 10 m.

$$mv^2 = 1 \times 20 \times 20 = 400 = m'v'^2 = 4 \times 10 \times 10 = 400.$$

On the other hand the times of fall are different : the 4 kg. traverse their 5 m. in 1 second, the 1 kg. traverses its 20 m. in 2 seconds. Friction and air resistance are, of course, neglected here.

But after each of the two bodies has fallen from its height, its motion ceases. Therefore, mv appears here as the measure of simple transferred, hence lasting, mechanical motion, and mv^2 as the measure of the vanished mechanical motion.

Further, the same thing applies to the impact of perfectly elastic bodies : the sum of both mv and of mv^2 is unaltered before and after impact. Both measures have the same validity.

This is not the case on impact of inelastic bodies.

Here, too, the current elementary textbooks (higher mechanics is hardly concerned at all with such trifles) teach that before and after impact the sum of mv remains the same. On the other hand a loss of *vis viva* occurs, for if the sum of mv^2 after impact is subtracted from the sum of mv^2 *before* impact, there is under all circumstances a positive remainder. By this amount (or the half of it, according to the notation adopted) the *vis viva* is diminished owing both to the mutual penetration and to the change of form of the colliding bodies. The latter is now clear and obvious, but not so the first assertion that the sum of mv remains the same before and after impact. In spite of Suter, *vis viva* is motion, and if a part of it is lost, motion is lost. Consequently, either mv here incorrectly expresses the quantity of motion, or the above assertion is untrue. In general the whole theorem has been handed down from a period when there was as yet no inkling of the transformation of motion; when, therefore, a disappearance of mechanical motion was only conceded where there was no other way out. Thus, the equality here of the sum of mv before and after impact was taken as proved by the fact that no loss or gain of this sum had been introduced. If, however, the bodies lose *vis viva* in internal friction corresponding to their inelasticity, they also lose velocity, and the sum of mv after impact must be smaller than before.[1] For it is surely not possible to neglect the internal friction in calculating mv, when it makes itself felt so clearly in calculating mv^2.

But this does not matter. Even if we admit the theorem, and calculate the velocity after falling, on the assumption that the sum of mv has remained the same, this decrease of the sum of mv^2 is still found.

---

[1] This is incorrect. Momentum remains constant even in an inelastic condition.

Here, therefore, mv and mv^2 conflict, and they do so by the difference of the mechanical motion that has actually disappeared. Moreover, the calculation itself shows that the sum of mv^2 expresses the quantity of motion correctly, while the sum of mv expresses it incorrectly.

Such are pretty nearly all the cases in which mv is employed in mechanics. Let us now glance at some cases in which mv^2 is employed.

When a cannon-ball is fired, it uses up in its course an amount of motion that is proportional to mv^2, irrespective of whether it encounters a solid target or comes to a standstill owing to air resistance and gravitation. If a railway train runs into a stationary one, the violence of the collision, and the corresponding destruction, is proportional to its mv^2. Similarly, mv^2 serves wherever it is necessary to calculate the mechanical force required for overcoming a resistance.

But what is the meaning of this convenient phrase, so current in mechanics : overcoming a resistance ?

If we overcome the resistance of gravity by raising a weight, there disappears a quantity of motion, a quantity of mechanical force, equal to that produced anew by the direct or indirect fall of the raised weight from the height reached back to its original level. The quantity is measured by half the product of the mass and the final velocity after falling, $\dfrac{mv^2}{2}$. What then occurred on raising the weight ? Mechanical motion, or force, disappeared as such. But it has not been annihilated ; it has been converted into mechanical force of tension, to use Helmholtz's expression ; into potential energy, as the moderns say ; into ergal as Clausius calls it ; and this can at any moment, by any mechanically appropriate means, be reconverted into the same quantity of mechanical motion as was necessary to produce it.

The potential energy is only the negative expression of the *vis viva* and *vice versa*.

A 24-lb. cannon-ball moving with a velocity of 400 m. per second strikes the one-metre thick armour-plating of a warship and under these conditions has apparently no effect on the armour. Consequently an amount of mechanical motion has vanished equal to $\frac{mv^2}{2}$, *i.e.* (since 24 lbs.=12 kg.) $=12 \times 400 \times 400 \times \frac{1}{2} = 960,000$ kilogram-metres. What has become of it? A small portion has been expended in the concussion and molecular alteration of the armour-plate. A second portion goes in smashing the cannon-ball into innumerable fragments. But the greater part has been converted into heat and raises the temperature of the cannon-ball to red heat. When the Prussians, in passing over to Alsen in 1864, brought their heavy batteries into play against the armoured sides of the Rolf Krake, after each hit they saw in the darkness the flare produced by the suddenly glowing shot. Even earlier, Whitworth had proved by experiment that explosive shells need no detonator when used against armoured warships; the glowing metal itself ignites the charge. Taking the mechanical equivalent of the unit of heat as 424 kilogram-metres, the quantity of heat corresponding to the above-mentioned amount of mechanical motion is 2,264 units. The specific heat of iron=0·1140; that is to say, the amount of heat that raises the temperature of 1 kg. of water by 1° C. (which serves as the unit of heat) suffices to raise the temperature of $\frac{1}{0\cdot1140} = 8\cdot772$ kg. of iron by 1° C. Therefore the 2,264 heat-units mentioned above raise the temperature of 1 kg. of iron by $8\cdot772 \times 2,264 = 19,860°$ C. or 19,860 kg. of iron by 1° C. Since this quantity of heat is distributed uniformly in the armour and the shot, the latter has its temperature raised by $\dfrac{19,860°}{2 \times 12} = 828°$, amounting to quite a good glowing heat. But since the foremost, striking end of the shot receives at any rate by far the greater part of the heat, certainly double that of the rear half, the former would be raised to a temperature of 1,104° C. and the latter to 552° C., which would fully suffice to explain the glowing effect even if we make a big deduction for the actual mechanical work performed on impact.

Mechanical motion also disappears in friction, to re-appear as heat; it is well known that, by the most accurate possible measurement of the two processes, Joule in Manchester and Colding in Copenhagen were the first to make an approximate experimental measurement of the mechanical equivalent of heat.

The same thing applies to the production of an electric current in a magneto-electrical machine by means of mechanical force, e.g. from a steam engine. The quantity of so-called electromotive force [1] produced in a given time is proportional to the quantity of mechanical motion used up in the same period, being equal to it if expressed in the same units. We can imagine this quantity of mechanical motion being produced, not by a steam engine, but by a weight falling in accordance with the pressure of gravity. The mechanical force that this is capable of supplying is measured by the *vis viva* that it would obtain on falling freely through the same distance, or by the force required to raise it again to the original height; in both cases $\dfrac{mv^2}{2}$.

Hence we find that while it is true that mechanical

---

[1] The phrase " electromotive force " is now used in a much stricter sense than 50 years ago. It is the quantity measured in volts. The quantity equivalent to mechanical energy is of course electrical energy measured in kilowatt-hours. These terms were only exactly defined and accurately measured after electrical energy became a commodity.

motion has a two-fold measure, each of these measures holds good for a very definitely demarcated series of phenomena. If already existing mechanical motion is transferred in such a way that it remains as mechanical motion, the transference takes place in proportion to the product of the mass and the velocity. If, however, it is transferred in such a way that it disappears as mechanical motion in order to reappear in the form of potential energy, heat, electricity, etc., in short, if it is converted into another form of motion, then the quantity of this new form of motion is proportional to the product of the originally moving mass and the square of the velocity. In short, mv is mechanical motion measured as mechanical motion ; $\dfrac{mv^2}{2}$ is mechanical motion measured by its capacity to become converted into a definite quantity of another form of motion. And, as we have seen, these two measures, because different, do not contradict one another.

It becomes clear from this that Leibniz's quarrel with the Cartesians was by no means a mere verbal dispute, and that d'Alembert's verdict in point of fact settled nothing at all. D'Alembert might have spared himself his tirades on the unclearness of his predecessors, for he was just as unclear as they were. In fact, as long as it was not known what becomes of the apparently annihilated mechanical motion, the absence of clarity was inevitable. And as long as mathematical mechanicians like Suter remain obstinately shut in by the four walls of their special science, they are bound to remain just as unclear as d'Alembert and to put us off with empty and contradictory phrases.

But how does modern mechanics express this conversion of mechanical motion into another form of motion, proportional in quantity to the former ? It has *performed work*, and indeed a definite amount of work.

But this does not exhaust the concept of work in the physical sense of the word. If, as in a steam or heat engine, heat is converted into mechanical motion, *i.e.* molecular motion is converted into mass motion, if heat breaks up a chemical compound, if it becomes converted into electricity in a thermopile, if an electric current sets free the elements of water from dilute sulphuric acid, or, conversely, if the motion (alias energy) produced in the chemical process of a current-producing cell takes the form of electricity and this in the circuit once more becomes converted into heat— in all these processes the form of motion that initiates the process, and which is converted by it into another form, performs work, and indeed a quantity of work corresponding to its own quantity.

Work, therefore, is change of form of motion regarded in its quantitative aspect.

But how so ? If a raised weight remains suspended and at rest, is its potential energy during the period of rest also a form of motion ? Certainly. Even Tait arrives at the conviction that potential energy is subsequently resolved into a form of actual motion [1] (*Nature*, XIV, p. 459). And, apart from that, Kirchhoff goes much further in saying (*Mathematical Mechanics*, p. 32) " Rest is a special case of motion," and thus proves that he can not only calculate but can also think dialectically.

Hence, by a consideration of the two measures of mechanical motion, we arrive incidentally, easily, and almost as a matter of course, at the concept of work, which was described to us as being so difficult to comprehend without mathematical mechanics. At any rate, we now know more about it than from Helmholtz's

---

[1] In Einstein's general theory of relativity space-time is distorted by a gravitational field, and therefore the relation between two bodies separated by a gravitational field is of the same character as if they were in relative motion. In this sense potential energy may be said to be resolved into motion.

lecture *On the Conservation of Force* (1862), which was
intended precisely " to make as clear as possible the
fundamental physical concepts of work and their invariability." All that we learn there about work is : that
it is something which is expressed in foot-pounds or in
units of heat, and that the number of these foot-pounds
or units of heat is invariable for a definite quantity of
work ; and, further, that besides mechanical forces
and heat, chemical and electric forces can perform work,
but that all these forces exhaust their capacity for work
in the measure that they actually result in work. We
learn also that it follows from this that the sum of all
effective quantities of force in nature as a whole remains
eternally and invariably the same throughout all the
changes taking place in nature. The concept of work is
neither developed, nor even defined.[1] And it is precisely the quantitative invariability of the magnitude
of work which prevents him from realising that the
qualitative alteration, the change of form, is the basic
condition for all physical work. And so Helmholtz can
go so far as to assert that " friction and inelastic impact
are processes in which *mechanical work is destroyed*
and heat is produced instead." (*Pop. Vorträge* [*Popular
Lectures*], II, p. 166.) Just the contrary. Here
mechanical work is not *destroyed*, here mechanical work
is *performed*. It is mechanical *motion* that is apparently
destroyed. But mechanical motion *can* never perform
even a millionth part of a kilogram-metre of work,
without apparently being destroyed as such, without
becoming converted into another form of motion.

But, as we have seen, the capacity for work contained
in a given quantity of mechanical motion is what is

---

[1] We get no further by consulting Clerk Maxwell. The latter says
(*Theory of Heat*, 4th edition, London, 1875, p. 87) : " Work is done
when resistance is overcome," and on p. 183, " The energy of a body is
its capacity for doing work." That is all that we learn about it.
[*Note by F. Engels.*]

known as its *vis viva*, and until recently was measured by mv^2. And here a new contradiction arose. Let us listen to Helmholtz (*Conservation of Force*, p. 9).

We read there that the magnitude of work can be expressed by a weight m being raised to a height h, when, if the force of gravity is put as g, the magnitude of work $=mgh$. For the body m to rise freely to the vertical height h, it requires a velocity $v=\sqrt{2gh}$, and it attains the same velocity on falling. Consequently, $mgh=\dfrac{mv^2}{2}$, and Helmholtz proposes " to take the magnitude $\dfrac{mv^2}{2}$ as the quantity of *vis viva*, whereby it becomes identical with the measure of the magnitude of work. From the viewpoint of how the concept of *vis viva* has been applied hitherto . . . this change has no significance, but it will offer essential advantages in the future."

It is scarcely to be believed. In 1847, Helmholtz was so little clear about the mutual relations of *vis viva* and work, that he totally fails to notice how he transforms the former proportional measure of *vis viva* into its absolute measure, and remains quite unconscious of the important discovery he has made by his audacious handling, recommending his $\dfrac{mv^2}{2}$ only because of its convenience as compared with mv^2 ! And it is as a matter of convenience that mechanicians have adopted $\dfrac{mv^2}{2}$. Only gradually was $\dfrac{mv^2}{2}$ also proved mathematically. Naumann (*Allg. Chemie* [*General Chemistry*], p. 7) gives an algebraical proof, Clausius (*Mechanische Wärmetheorie* [*The Mechanical Theory of Heat*], 2nd edition, p. 18), an analytical one, which is then to be met with in another form and a different method of deduction in Kirchhoff (*ibid.*, p. 27).

Clerk Maxwell (*ibid.*, p. 88)[1] gives an elegant algebraical proof of the deduction of $\frac{mv^2}{2}$ from mv. This
does not prevent our two Scotsmen, Thomson and Tait,
from asserting (*ibid.*, p. 163) : " The *vis viva* or kinetic
energy of a moving body is proportional to the mass
and the square of the velocity conjointly. If we adopt
the same units of mass as above (namely, unit of mass
moving with unit velocity) there is a *particular advantage*
in defining kinetic energy as *half* the product of the
mass and the square of the velocity." Here, therefore,
we find that not only the ability to think, but also to
calculate, has come to a standstill in the two foremost
mechanicians of Scotland. The particular advantage,
the convenience of the formula, accomplishes everything
in the most beautiful fashion.

For us, who have seen that *vis viva* is nothing but the
capacity of a given quantity of mechanical motion to
perform work, it is obvious on the face of it that the
expression in mechanical terms of this capacity for work
and the work actually performed by the latter must
be equal to each other ; and that, consequently, if $\frac{mv^2}{2}$
measures the work, the *vis viva* must likewise be
measured by $\frac{mv^2}{2}$. But that is what happens in science.

Theoretical mechanics arrives at the concept of *vis viva*,
the practical mechanics of the engineer arrives at the
concept of work and forces it on the theoreticians.[2]
And, immersed in their calculations, the theoreticians
have become so unaccustomed to thinking that for years

[1] See Appendix II, p. 332.
[2] The term *vis viva*, measured by mv^2, has now completely disappeared from theoretical mechanics, as Engels thought that it should.
To-day most people can think in terms of energy, not because of any
theoretical advances, but because it is a commodity. We buy it in
therms, calories, kilowatt-hours, and other measures, and are therefore
forced to think about it in a concrete manner.

they fail to recognise the connection between the two
concepts, measuring one of them by mv^2, the other by
$\dfrac{mv^2}{2}$, and finally accepting $\dfrac{mv^2}{2}$ for both, not from comprehension, but for the sake of simplicity of calculation ! [1]

[1] The word " work " and the corresponding idea is derived from
English engineers. But in English practical work is called " work,"
while work in the economic sense is called " labour." Hence, physical
work also is termed " work," thereby excluding all confusion with
work in the economic sense. This is not the case in German ; therefore
it has been possible in recent pseudo-scientific literature to make
various peculiar applications of work in the physical sense to economic
conditions of labour and vice versa. But we have also the word " Werk "
which, like the English word " work," is excellently adapted for
signifying physical work. Economics, however, being a sphere far too
remote from our natural scientists, they will scarcely decide to introduce
it to replace the word Arbeit, which has already obtained general
currency—unless, perhaps, when it is too late. Only Clausius * has
made the attempt to retain the expression " Werk," at least alongside
the expression " Arbeit." [Note by F. Engels.]

* See Appendix II, p. 333.

V

HEAT

As we have seen, there are two forms in which mechanical motion, *vis viva*, disappears. The first is its conversion into mechanical potential energy, for instance on lifting a weight. This form has the peculiarity that not only can it be re-transformed into mechanical motion—this mechanical motion, moreover, having the same *vis viva* as the original one—but also that it is capable only of this change of form. Mechanical potential energy can never produce heat or electricity, unless it has been converted first into real mechanical motion. To use Clausius' term, it is a " reversible process."

The second form in which mechanical motion disappears is in friction and impact—which differ only in degree. Friction can be conceived as a series of small impacts occurring successively and side by side, impact as friction concentrated at one spot and in a single moment of time. Friction is chronic impact, impact is acute friction. The mechanical motion that disappears here, disappears altogether *as such*. It can never be restored immediately out of itself. The process is not directly reversible. The motion has been transformed into qualitatively different forms of motion, into heat, electricity—into forms of molecular motion.

Hence, friction and impact lead from the motion of masses, the subject matter of mechanics, to molecular motion, the subject matter of physics.

In calling physics the mechanics of molecular motion, it has not been overlooked that this expression by no

means covers the entire field of contemporary physics.
On the contrary. Ether vibrations, which are responsible
for the phenomena of light and radiant heat, are certainly
not molecular motions in the modern sense of the word.
But their terrestrial actions concern molecules first and
foremost : refraction of light, polarisation of light, etc.,
are determined by the molecular constitution of the
bodies concerned. Similarly almost all the most important scientists now [1] regard electricity as a motion
of ether particles, and Clausius even says of heat [2]
that in " the movement of ponderable atoms (it would
be better to say molecules) . . . the ether within the
body can also participate " (*Mechanische Wärmetheorie*
[*Mechanical Theory of Heat*] I, p. 22).[3] But in the
phenomena of electricity and heat, once again it is
primarily molecular motions that have to be considered ;
it could not be otherwise, so long as our knowledge of
the ether is so small. But when we have got so far as
to be able to present the mechanics of the ether, this
subject will include a great deal that is now of necessity
allocated to physics.[4]

The physical processes in which the structure of the molecule is altered, or even destroyed, will be dealt with later
on : they form the transition from physics to chemistry.

Only with molecular motion does the change of form
of motion acquire complete freedom. Whereas, at the
boundary of mechanics the motion of masses can assume

[1] At this time the ideas of Faraday and Maxwell were dominant,
and physicists tended to regard electricity as primarily located in the
field between charged bodies.

[2] A body at any temperature is in equilibrium with a certain density
of radiation, though very little of the energy in a given volume is
" in the ether," *i.e.* in the form of radiation, at ordinary temperatures.

[3] See Appendix II, pp. 333-4.

[4] This has certainly been verified in the sense that for modern
physics the properties of particles can be regarded as essentially repulsions and attractions in the space around them, which is also full of
radiation. On the other hand, the idea of the ether has proved so full
of internal contradictions that the word is now little used.

only a few other forms—heat or electricity—here, a quite different and more lively capacity for change of form is to be seen. Heat passes into electricity in the thermopile, it becomes identical [1] with light at a certain stage of radiation, and in its turn reproduces mechanical motion. Electricity and magnetism, a twin pair like heat and light, not only become transformed into each other, but also into heat and light as well as mechanical motion. And this takes place in such definite measure relations that a given quantity of any one of these forms of energy can be expressed in any other—in kilogram-metres, in heat units, in volts,[2] and similarly any unit of measurement can be translated into any other.

The practical discovery of the conversion of mechanical motion into heat is so very ancient that it can be taken as dating from the beginning of human history.[3] Whatever discoveries, in the way of tools and domestication of animals,[4] may have preceded it, the making of fire by friction was the first instance of men pressing a non-living force of nature into their service. Popular superstitions to-day still show how greatly the almost immeasurable import of this gigantic advance impressed itself on the mind of mankind. Long after the introduction of the use of bronze and iron the discovery of the stone knife, the first tool, continued to be celebrated, all religious sacrifices being performed with stone knives. According to the Jewish legend, Joshua decreed that men born in the wilderness should be circumcised with

[1] As we saw, some of the heat in a hot body takes the form of radiation. When the body gets red hot this becomes partially visible (*i.e.* light).

[2] This is, of course, a mistake. The volt is not an energy unit, as Engels would soon have known had he ever had to pay an electricity bill!

[3] Even *Sinanthropus*, a type of man very different physically from ourselves, possessed fire, though of course we do not know how he made it.

[4] The use of fire immensely preceded domestication.

stone knives ; the Celts and Germans used stone knives exclusively in their human sacrifices.   But all this long ago passed into oblivion.   It was different with the making of fire by friction.   Long after other methods of producing fire had become known, every sacred fire among the majority of peoples had to be obtained by friction.   But even to-day, popular superstition in the majority of the European countries insists that fire with miraculous powers (*e.g.* our German bonfire against epidemics) may be lighted only by means of friction.   Thus, down to our own day, the grateful memory of the first great victory of mankind over nature lives on—half unconsciously—in popular superstition, in the relics of heathen-mythological recollections, among the most educated peoples in the world.

However, the process of making fire by friction is still one-sided.   By it mechanical motion is converted into heat.   To complete the process, it must be reversed ; heat must be converted into mechanical motion.   Only in that case is justice done to the dialectics of the process, the cycle of the process being completed—for the first stage, at least.   But history has its own pace, and however dialectical its course may be in the last analysis, dialectics has often to wait for history a fairly long time. Many thousands of years must have elapsed between the discovery of fire by friction and the time when Hero of Alexandria (*ca.* 120 B.C.) invented a machine which was set in rotary motion by the steam issuing from it.   And almost another two thousand years elapsed before the first steam engine was built, the first apparatus for the conversion of heat into really useable mechanical motion.

The steam engine was the first really international invention, and this fact, in turn, testifies to a mighty historical advance.   The Frenchman, Papin, invented the first steam engine, and he invented it in Germany.

It was the German, Leibniz, scattering around him, as always, brilliant ideas, without caring whether the merit for them would be awarded to him or someone else, who, as we know now from Papin's correspondence (published by Gerland), gave him the main idea of the machine : the employment of a cylinder and piston. Soon after that, the Englishmen, Savery and Newcomen, invented similar machines ; finally, their fellow-countryman, Watt, by introducing a separate condenser, brought the steam engine in principle up to the level of to-day.[1] The cycle of inventions in this sphere was completed ; the conversion of heat into mechanical motion was achieved. What came afterwards were improvements in details.

Practice, therefore, solved after its own fashion the problem of the relations between mechanical motion and heat. It had, to begin with, converted the first into the second, and then it converted the second into the first. But how did matters stand in regard to theory ?

The situation was pitiable enough. Although it was just in the seventeenth and eighteenth centuries that innumerable accounts of travel appeared, teeming with descriptions of savages who knew no way of producing fire other than by friction, yet physicists were almost uninterested in it ; they were equally indifferent to the steam engine during the whole of the eighteenth century and the first decades of the nineteenth. For the most part they were satisfied simply to record the facts.

Finally, in the 'twenties, Sadi Carnot took the matter in hand, and indeed so very skilfully that his best calculations, afterwards presented by Clapeyron in geometrical form, have been accepted up to the present day by Clausius and Clerk Maxwell. Sadi Carnot almost got to the bottom of the question. It was not the lack of factual data that prevented him from completely

[1] The turbine was of course only introduced in 1884.

solving it, but solely a preconceived *false theory*. Moreover, this false theory was not one which had been forced upon physicists by some variety of malicious philosophy, but was a theory contrived by the physicists themselves, by means of their own naturalistic mode of thought, so very superior to the metaphysical-philosophical method.

In the seventeenth century heat was regarded, at any rate in England, as a property of bodies, as " a *motion* of a particular kind, the nature of which has never been explained in a satisfactory manner." This is what Th. Thomson called it, two years before the discovery of the mechanical theory of heat (*Outline of the Sciences of Heat and Electricity*, 2nd edition, London, 1840). But in the eighteenth century the view came more and more to the fore that heat, as also light, electricity, and magnetism, is a special substance, and that all these peculiar substances differ from ordinary matter in having no weight, in being imponderable.

VI

ELECTRICITY[1]

ELECTRICITY, like heat, only in a different way, has also a certain omnipresent character. Hardly any change can occur in the world without it being possible to demonstrate the presence of electrical phenomena. If water evaporates, if a flame burns, if two different metals, or two metals of different temperature, touch, or if iron touches a solution of copper sulphate, and so on, electrical processes take place simultaneously with the more apparent physical and chemical phenomena. The more exactly we investigate natural processes of the most diverse nature, the more do we find evidence of electricity. In spite of its omnipresence, in spite of the fact that for half a century electricity has become more and more pressed into the industrial service of mankind, it remains precisely that form of motion the nature of which is still enveloped in the greatest obscurity.

The discovery of the galvanic current is approximately 25 years younger than that of oxygen and is at least as significant for the theory of electricity as the latter discovery was for chemistry. Yet what a difference obtains even to-day between the two fields! In chemistry, thanks especially to Dalton's discovery of

[1] For the factual material in this chapter we rely mainly on Wiedemann's *Lehre vom Galvanismus und Elektromagnetismus* [*Theory of Galvanism and Electro-Magnetism*], 2 vols. in 3 parts, 2nd edition, Braunschweig, 1874.

In *Nature*, June 15, 1882, there is a reference to this " admirable treatise, which in its forthcoming shape, with electrostatics added, will be the greatest experimental treatise on electricity in existence." [*Note by F. Engels.*]

atomic weights, there is order, relative certainty about what has been achieved, and systematic, almost planned, attack on the territory still unconquered, comparable to the regular siege of a fortress. In the theory of electricity there is a barren lumber of ancient, doubtful experiments, neither definitely confirmed nor definitely refuted ; an uncertain fumbling in the dark, unco-ordinated research and experiment on the part of numerous isolated individuals, who attack the unknown territory with their scattered forces like the attack of a swarm of nomadic horsemen. It must be admitted, indeed, that in the sphere of electricity a discovery like that of Dalton, giving the whole science a central point and a firm basis for research, is still to seek.[1] It is essentially this unsettled state of the theory of electricity, which for the time being makes it impossible to establish a comprehensive theory, that is responsible for the fact that a one-sided empiricism prevails in this sphere, an empiricism which as far as possible itself forbids thought, and which precisely for that reason not only thinks incorrectly but also is incapable of faithfully pursuing the facts or even of reporting them faithfully, and which, therefore, becomes transformed into the opposite of true empiricism.

If in general those natural scientists, who cannot say anything bad enough of the crazy *a priori* speculations of the German philosophy of nature, are to be recommended to read the theoretico-physical works of the empirical school, not only of the contemporary but even of a much later period, this holds good quite especially for the theory of electricity. Let us take a work of the year 1840 : *An Outline of the Sciences of Heat and Electricity*, by Thomas Thomson. Old Thomson was indeed an authority in his day ; moreover he had already at his disposal a very considerable part of

[1] The central discovery was J. J. Thomson's discovery of the electron.

the work of the greatest electrician so far—Faraday.
Yet his book contains at least just as crazy things as the
corresponding section of the much older Hegelian philosophy of nature. The description of the electric spark,
for instance, might have been translated directly from
the corresponding passage in Hegel. Both enumerate
all the wonders that people sought to discover in the
electric spark, prior to knowledge of its real nature and
manifold diversity, and which have now been shown to
be mainly special cases or errors.

Still better, Thomson recounts quite seriously on p. 446
Dessaigne's cock-and-bull stories, such as that, with a
rising barometer and falling thermometer, glass, resin,
silk, etc., become negatively electrified on immersion in
mercury, but positively if instead the barometer is
falling and the temperature rising ; that in summer
gold and several other metals become positive on warming
and negative on cooling, but in winter the reverse ; that
with a high barometer and northerly wind they are
strongly electric, positive if the temperature is rising and
negative if it is falling, etc.

So much for the treatment of the facts. As regards
a priori speculation, Thomson favours us with the
following treatment of the electric spark, derived from
no lesser person than Faraday himself :

" The spark is a discharge . . . or weakening of the
polarised inductive state of many dielectric particles
by means of a peculiar action of a few of these particles
occupying a very small and limited space. Faraday
assumes that the few particles situated where the
discharge occurs are not merely pushed apart, but
assume a peculiar, highly exalted, condition for the
time, *i.e.* that they have thrown on them all the
surrounding forces in succession and are thus brought
into a proportionate intensity of condition, perhaps
equal to that of chemically combining atoms ; that
they then discharge the powers, in the same manner

as the atoms do their, in some way at present unknown to us and so the end of the whole. The
ultimate effect is exactly as if a metallic wire had been
put into the place of the discharging particles, and
it does not seem impossible that the principles of
action in both cases may, hereafter, prove to be the
same." [1]

I have, adds Thomson, given this explanation of
Faraday's in his own words, because I do not understand
it clearly. This will certainly have been the experience
of other persons also, quite as much as when they read
in Hegel that in the electric spark " the special materiality of the charged body does not as yet enter into the
process but is determined within it only in an elementary and spiritual way," and that electricity is " the
anger, the effervescence, proper to the body," its " angry
self " that " is exhibited by every body when excited."
(*Philosophy of Nature*, paragraph 324, addendum.) [2]

Yet the basic thought of both Hegel and Faraday is
the same. Both oppose the idea that electricity is not
a state of matter but a special, distinct variety of matter.
And since in the spark electricity is apparently exhibited
as independent, free from any foreign material substratum, separated out and yet perceptible to the senses,
they arrive at the necessity, in the state of science at
the time, of having to conceive of the spark as the
transient phenomenal form of a " force " momentarily
freed from all matter. For us, of course, the riddle is
solved, since we know that on the spark discharge
between metal electrodes real " metallic particles "
leap across, and hence in actual fact " the special
materiality of the charged body enters into the process."

As is well known, electricity and magnetism, like
heat and light, were at first regarded as special imponderable substances. As far as electricity is concerned,

[1,2] See Appendix II, p. 334.

it is well known that the view soon developed that there
are two opposing substances, two " fluids," one positive
and one negative, which in the normal state neutralise
each other, until they are forced apart by a so-called
" electric force of separation." It is then possible to
charge two bodies, one with positive, the other with
negative electricity ; on uniting them by a third conducting body equalisation occurs, either suddenly or
by means of a lasting current, according to circumstances. The sudden equalisation appeared very simple
and comprehensible, but the current offered difficulties.
The simplest hypothesis, that the current in every case
is a movement of either purely positive or purely
negative electricity, was opposed by Fechner, and in
more detail by Weber, with the view that in every circuit
two equal currents of positive and negative electricity
flow in opposite directions in channels lying side by side
between the ponderable molecules of the bodies.[1]
Weber's detailed mathematical working out of this
theory finally arrives at the result that a function, of
no interest to us here, is multiplied by a magnitude l/r,
the latter signifying " *the ratio . . . of the unit of
electricity to the milligram.*" (Wiedemann, *Lehre vom
Galvanismus, etc.* [*Theory of Galvanism, etc.*], 2nd edition,
III, p. 569). The ratio to a measure of weight can
naturally only be a weight ratio. Hence one-sided
empiricism had already to such an extent forgotten the
practice of thought in calculating that here it even makes
the imponderable electricity ponderable and introduces
its weight into the mathematical calculation.

The formulæ derived by Weber sufficed only within
certain limits, and Helmholtz, in particular, only a few
years ago calculated results that come into conflict

---

[1] We now know that a current in metals is due to a movement of
electrons, whereas in electrolytes, *e.g.* salt water and gases, molecules
with both positive and negative charges carry it.

with the principle of the conservation of energy. In opposition to Weber's hypothesis of the double current flowing in opposite directions, C. Naumann in 1871 put forward the other hypothesis that in the current only one of the two electricities, for instance the positive, moves, while the other negative one remains firmly bound up with the mass of the body. On this Wiedemann includes the remark : " This hypothesis could be linked up with that of Weber if to Weber's supposed double current of electric masses $\pm\frac{1}{2}e$ flowing in opposite directions, there were added a further *current of neutral electricity*, externally inactive, which carried with it amounts of electricity $\pm\frac{1}{2}e$ in the direction of the positive current." (III, p. 577.)

This statement is once again characteristic of one-sided empiricism. In order to bring about the flow of electricity at all, it is decomposed into positive and negative. All attempts, however, to explain the current with these two substances, meet with difficulties ; both the assumption that only one of them is present in the current and that the two of them flow in opposite directions simultaneously, and, finally, the third assumption also that one flows and the other is at rest. If we adopt this last assumption how are we to explain the inexplicable idea that negative electricity, which is mobile enough in the electrostatic machine and the Leyden jar, in the current is firmly united with the mass of the body ? Quite simply. Besides the positive current $+e$, flowing through the wire to the right, and the negative current, $-e$, flowing to the left, we make yet another current, this time of neutral electricity, $\pm\frac{1}{2}e$, flow to the right. First we assume that the two electricities, to be able to flow at all, must be separated from one another ; and then, in order to explain the phenomena that occur on the flow of the separated electricities, we assume that they can also flow unseparated. First

we make a supposition to explain a particular phenomenon, and at the first difficulty encountered we make
a second supposition which directly negates the first one.
What must be the sort of philosophy that these gentlemen have the right to complain of ?

However, alongside this view of the material nature of
electricity, there soon appeared a second view, according
to which it is to be regarded as a mere state of the body,
a " force " or, as we would say to-day, a special form of
motion. We saw above that Hegel, and later Faraday,
adhered to this view. After the discovery of the
mechanical equivalent of heat had finally disposed of
the idea of a special " heat stuff," and heat was shown
to be a molecular motion, the next step was to treat
electricity also according to the new method and to
attempt to determine its mechanical equivalent. This
attempt was fully successful. Particularly owing to
the experiments of Joule, Favre, and Raoult, not only
was the mechanical and thermal equivalent of the so-
called " electromotive force " of the galvanic current
established, but also its complete equivalence with the
energy liberated by chemical processes in the exciting
cell or used up in the decomposition cell. This made the
assumption that electricity is a special material fluid
more and more untenable.

The analogy, however, between heat and electricity
was not perfect. The galvanic currents still differed in
very essential respects from the conduction of heat.
It was still not possible to say *what* it was that moved
in the electrically affected bodies. The assumption of a
mere molecular vibration as in the case of heat seemed
insufficient. In view of the enormous velocity of
motion of electricity, even exceeding that of light,[1]
it remained difficult to overcome the view that here some

[1] This is incorrect, but was generally stated in textbooks at the
time when Engels wrote.

material substance is in motion between the molecules of the body.

Here the most recent theories put forward by Clerk Maxwell (1864), Hankel (1865), Reynard (1870), and Edlund (1872) are in complete agreement with the assumption already advanced in 1846, first of all as a suggestion by Faraday, that electricity is a movement of the elastic medium permeating the whole of space and hence all bodies as well, the discrete particles of which medium repel one another according to the law of the inverse square of the distance. In other words, it is a motion of ether particles, and the molecules of the body take part in this motion. As to the manner of this motion, the various theories are divergent; those of Maxwell, Hankel, and Reynard, taking as their basis modern investigations of vortex motion, explain it in various ways from vortices, so that the vortex of old Descartes also once more comes into favour in an increasing number of new fields. We refrain from going more closely into the details of these theories. They differ strongly from one another and they will certainly still experience many transformations. But a decisive advance appears to lie in their common basic conception : that electricity is a motion of the particles of the luminiferous ether that penetrates all ponderable matter, this motion reacting on the molecules of the body. This conception reconciles the two earlier ones. According to it, it is true that in electrical phenomena it is something substantial that moves, something different from ponderable matter. But this substance is not electricity itself, which in fact proves rather to be a form of motion, although not a form of the imme-diate direct motion of ponderable matter. While, on the one hand, the ether theory shows a way of getting over the primitive clumsy idea of two opposed electrical fluids, on the other hand it gives a prospect of explaining *what* the real, substantial substratum of electrical motion is, *what* sort of a thing it is whose motion produces electrical phenomena.[1]

The ether theory has already had *one* decisive success. As is well known, there is at least one point where electricity directly alters the motion of light : it rotates the latter's plane of polarisation. On the basis of his theory mentioned above, Clerk Maxwell calculates that the electric specific inductive capacity of a body is equal to the square of its index of refraction. Boltzmann has investigated dielectric coefficients of various non-conductors and he found that in sulphur, rosin, and paraffin, the square roots of these coefficients were respectively equal to their indices of refraction. The highest deviation—in sulphur—amounted to only 4 per cent. Consequently, the Maxwellian ether theory in this particular has hereby been experimentally confirmed.[2]

It will, however, require a lengthy period and cost much labour before new series of experiments will have extracted a firm kernel from these mutually contradictory hypotheses. Until then, or until the ether theory, too, is perhaps supplanted by an entirely new one, the theory of electricity finds itself in the uncomfortable position of having to employ a mode of expression which it itself admits to be false. Its whole terminology is still based on the idea of two electric fluids. It still speaks quite unashamedly of " electric masses flowing in the bodies," of " a division of electricities in every molecule," etc. This is a misfortune which for the most part, as already said, follows in-

---

[1] The view that electrical energy was located in the ether was the basis of the experiments which gave us radio. It seemed in turn to have been negated by the discovery of electrons. However, the electron in turn is now regarded by many physicists as a system of waves rather than a well-defined particle.

[2] Every broadcast is a confirmation of this theory to-day.

evitably from the present transitional state of science, but which also, with the one-sided empiricism particularly prevalent in this branch of investigation, contributes not a little to preserving the existing confusion of thought.

The opposition between so-called static or frictional electricity and dynamic electricity or galvanism can now be regarded as bridged over, since we have learned to produce constant currents by means of the electric machine and, conversely, by means of the galvanic current to produce so-called static electricity, to charge Leyden jars, etc. We shall not here touch on the sub-form of static electricity, nor likewise on magnetism, which is now recognised to be also a sub-form of electricity. The theoretical explanation of the phenomena belonging here will under all circumstances have to be sought in the theory of the galvanic current, and consequently we shall keep mainly to this.

A constant current can be produced in many different ways. Mechanical mass motion produces *directly*, by friction, in the first place only static electricity, and a constant current only with great dissipation of energy. For the major part, at least, to become transformed into electric motion, the intervention of magnetism is required, as in the well-known magneto-electric machines[1] of Gramme, Siemens, and others. Heat can be converted directly into current electricity, as especially occurs at the junction of two different metals. The energy set free by chemical action, which under ordinary circumstances appears in the form of heat, is converted under appropriate conditions into electric motion. Cor versely, the latter form of motion, as soon as the requisite conditions are present, passes into any other form of motion : into mass motion, to a very small extent directly into electro-dynamic attractions and repulsions ; to a large extent, however, by the inter-

[1] Now called dynamos.

vention of magnetism in the electro-magnetic machine ; into heat—throughout a closed circuit, unless other changes are brought about ; into chemical energy— in decomposition cells and voltameters introduced into the circuit, where the current dissociates compounds that are attacked in vain by other means.

All these transformations are governed by the basic law of the quantitative equivalence of motion through all its changes of form. Or, as Wiedemann expresses it : " By the law of conservation of force the mechanical work exerted in any way for the production of the current must be equivalent to the work exerted in producing all the effects of the current." The conversion of mass motion or heat into electricity [1] offers us no difficulties here ; it has been shown that the so-called " electromotive force " [2] in the first case is equal to the work expended on that motion, and in the second case it is " at every junction of the thermopile directly proportional to its absolute temperature " (Wiedemann, III, p. 482), *i.e.* to the quantity of heat present at every junction measured in absolute units. The same law has in fact been proved valid also for electricity produced from chemical energy. But here the matter seems to be not so simple, at least for the theory now current. Let us, therefore, go into this somewhat more deeply.

One of the most beautiful series of experiments on the transformations of form of motion as a result of the action of a galvanic cell is that of Favre (1857–58). He put a Smee cell of five elements in a calorimeter ;

[1] I use the term " electricity " in the sense of electric motion with the same justification that the general term " heat " is used to express the form of motion that our senses perceive as heat. This is the less open to objection in as much as any possible confusion with the state of *stress* of electricity is here expressly excluded in advance. [*Note by F. Engels.*]

[2] Once more it must be remembered that this term was very loosely used sixty years ago, and now has a definite meaning, not of course equivalent to any form of energy.

in a second calorimeter he put a small electro-magnetic motor, with the main axle and driving wheel projecting so as to be available for any kind of coupling. Each production in the cell of one gram of hydrogen, or solution of 32·6 grams of zinc (the old chemical equivalent of zinc, equal to half the now accepted atomic weight 65·2, and expressed in grams), gave the following results :

A. The cell enclosed in the calorimeter, excluding the motor : heat production 18,682 or 18,674 units of heat.

B. Cell and motor linked in the circuit, but the motor prevented from moving : heat in the cell 16,448, in the motor 2,219, together 18,667 units of heat.

C. As B, but the motor in motion without however lifting a weight : heat in the cell 13,888, in the motor 4,769, together 18,657 units of heat.

D. As C, but the motor raises a weight and so performs mechanical work=131,24 kilogram-metres : heat in the cell 15,427, in the motor 2,947, total 18,374 units of heat ; loss in contrast to the above 18,682 equals 308 units of heat. But the mechanical work performed amounting to 131,24 kilogram-metres, multiplied by 1,000 (in order to bring the kilograms into line with the grams of the chemical results) and divided by the mechanical equivalent of heat= 423,5 kilogram-metres, gives 309 units of heat, hence exactly the loss mentioned above as the heat equivalent of the mechanical work performed.

The equivalence of motion in all its transformations is, therefore, strikingly proved for electric motion also, within the limits of unavoidable error. And it is likewise proved that the " electromotive force " of the galvanic battery is nothing but chemical energy converted into electricity, and the battery itself nothing but an apparatus that converts chemical energy on its

liberation into electricity, just as a steam engine transforms the heat supplied to it into mechanical motion,
without in either case the converting apparatus supplying
further energy on its own account.

A difficulty arises here, however, in relation to the
traditional mode of conception. The latter ascribes an
" *electric force of separation* " to the battery in virtue of
the conditions of contact present in it between the fluids
and metals, which force is proportional to the electromotive force and therefore for a given battery represents
a definite quantity of energy. What then is the relation
of this electric force of separation, which according to
the traditional mode of conception of the battery as such
is inherently a source of energy even without chemical
action, to the energy set free by chemical action ?
And if it is a source of energy independent of the latter,
whence comes the energy furnished by it ?

This question in a more or less unclear form constitutes the point of dispute between the contact theory
founded by Volta and the chemical theory of the galvanic
current that arose immediately afterwards.

The contact theory explained the current from the
electric stresses arising in the battery on contact of the
metals with one or more of the liquids, or even merely
on contact of the liquids themselves, and from their
neutralisation or that of the opposing electricities thus
generated in the circuit. The pure contact theory
regarded any chemical changes that might thereby
occur as quite secondary. On the other hand, as
early as 1805, Ritter maintained that a current could
only be formed if the excitants reacted chemically
even *before* closing the circuit. In general this older
chemical theory is summarised by Wiedemann (I,
p. 284) to the effect that according to it so-called contact
electricity " makes its appearance only if at the same
time there comes into play a real chemical action of the

bodies in contact, or at any rate a disturbance of the chemical equilibrium, even if not directly bound up with chemical processes, a 'tendency towards chemical action' between the bodies in contact."

It is seen that both sides put the question of the source of energy of the current only indirectly, as indeed could hardly be otherwise at the time. Volta and his successors found it quite in order that the mere contact of heterogeneous bodies should produce a constant current, and consequently be able to perform definite work without equivalent return. Ritter and his supporters are just as little clear how the chemical action makes the battery capable of producing the current and its performance of work. But if this point has long ago been cleared up for chemical theory by Joule, Favre, Raoult, and others, the opposite is the case for the contact theory. In so far as it has persisted, it remains essentially at the point where it started. Notions belonging to a period long outlived, a period when one had to be satisfied to ascribe a particular effect to the first available apparent cause that showed itself on the surface, regardless of whether motion was thereby made to arise out of nothing—notions that directly contradict the principle of the conservation of energy—thus continue to exist in the theory of electricity of to-day. And if the objectionable aspects of these ideas are shorn off, weakened, watered down, castrated, glossed over, this does not improve matters at all : the confusion is bound to become only so much the worse.

As we have seen, even the older chemical theory of the current declares the contact relations of the battery to be absolutely indispensable for the formation of the current : it maintains only that these contacts can never achieve a constant current without simultaneous chemical action. And even to-day it is still taken as a matter of course that the contact arrangements of the battery provide precisely the apparatus by means of which liberated chemical energy is transformed into electricity, and that it depends essentially on these contact arrangements whether and how much chemical energy actually passes into electric motion.

Wiedemann, as a one-sided empiricist, seeks to save what can be saved of the old contact theory. Let us follow what he has to say. He declares (I, p. 799) :

" In contrast to what was formerly believed, the effect of contact of chemically indifferent bodies, *e.g.* of metals, is *neither indispensable for the theory of the pile*, nor proved by the facts that *Ohm* derived his law from it, a law that can be derived without this assumption, and that *Fechner*, who confirmed this law experimentally, likewise defended the contact theory. Nevertheless, the excitation of electricity by *metallic* contact, according to the experiments now available at least, is not to be denied, even though the quantitative results obtainable in this respect may always be tainted with an inevitable uncertainty owing to the impossibility of keeping absolutely clean the surfaces of the bodies in contact."

It is seen that the contact theory has become very modest. It concedes that it is not at all indispensable for explaining the current, and neither proved theoretically by Ohm nor experimentally by Fechner. It even concedes then that the so-called fundamental experiments, on which alone it can still rest, can never furnish other than uncertain results in a quantitative respect, and finally it asks us merely to recognise that in general it is by contact—although only of *metals* !—that electric motion occurs.

If the contact theory remained content with this, there would not be a word to say against it. It will certainly be granted that on the contact of two metals electrical phenomena occur, in virtue of which a preparation of a frog's leg can be made to twitch, an electroscope charged, and other movements brought about. The only question that arises in the first place is : whence comes the energy required for this ?

To answer this question, we shall, according to Wiedemann (I, p. 14) :

" adduce *more or less the following* considerations : if the heterogeneous metal plates A and B are brought within a close distance of each other, they attract each other in consequence of the forces of adhesion. On mutual contact they lose the *vis viva* [1] of motion imparted to them by this attraction. (If we assume that the molecules of the metals are in a state of permanent vibration, it *could* also happen that, if on contact of the heterogeneous metals the molecules not vibrating simultaneously come into contact, an alteration of their vibration is thereby brought about with loss of *vis viva*.) The lost *vis viva* is *to a large extent* converted into heat. A *small portion* of it, however, is expended in bringing about a different distribution of the electricities previously unseparated. As we have already mentioned above, the bodies brought together become charged with equal quantities of positive and negative electricity, *possibly* as the result of an unequal attraction for the two electricities."

The modesty of the contact theory becomes greater and greater. At first it is admitted that the powerful electric force of separation, which has later such a gigantic work to perform, in itself possesses no energy of its own, and that it cannot function if energy is not supplied to it from outside. And then it has allotted to it a more than diminutive source of energy, the *vis viva* of adhesion, which only comes into play at scarcely measurable distances and which allows the bodies to travel a scarcely measurable length. But it does not matter : it indisputably exists and equally undeniably

[1] *I.e.* kinetic energy.

vanishes on contact. But even this minute source still furnishes too much energy for our purpose : a *large* part is converted into heat and only a *small* portion serves to evoke the electric force of separation. Now, although it is well known that cases enough occur in nature where extremely minute impulses bring about extremely powerful effects, Wiedemann himself seems to feel that his hardly trickling source of energy can with difficulty suffice here, and he seeks a possible second source in the assumption of an interference of the molecular vibrations of the two metals at the surfaces of contact. Apart from other difficulties encountered here, Grove and Gassiot have shown that for exciting electricity actual contact is not at all indispensable, as Wiedemann himself tells us on the previous page. In short, the more we examine it the more does the source of energy for the electric force of separation dwindle to nothing.

Yet up to now we hardly know of any other source for the excitation of electricity on metallic contact. According to Naumann (*Allg. u. phys. Chemie* [*General and Physical Chemistry*], Heidelberg, 1877, p. 675), " the contact-electromotive forces convert heat into electricity " ; he finds " the assumption natural that the ability of these forces to produce electric motion depends on the quantity of heat present, or, in other words, that it is a function of the temperature," as has also been proved experimentally by Le Roux. Here too we find ourselves groping in the dark. The law of the voltaic series of metals forbids us to have recourse to the chemical processes that to a small extent are continually taking place at the contact surfaces, which are always covered by a thin layer of air and impure water, a layer as good as inseparable as far as we are concerned. An electrolyte should produce a constant current in the circuit, but the electricity of mere metallic contact, on the contrary, disappears on closing the circuit. And here we come to the real point : whether, and in what manner, the production of a constant current on the contact of chemically indifferent bodies is made possible by this " electric force of separation," which Wiedemann himself first of all restricted to metals, declaring it incapable of functioning without energy being supplied from outside, and then referred exclusively to a truly microscopical source of energy.

The voltaic series arranges the metals in such a sequence that each one behaves as electro-negative in relation to the preceding one and as electro-positive in relation to the one that follows it.   Hence if we arrange a series of pieces of metal in this order, *e.g.* zinc, tin, iron, copper, platinum, we shall be able to obtain differences of electric potential at the two ends.   If, however, we arrange the series of metals to form a circuit so that the zinc and platinum are in contact, the electric stress is at once neutralised and disappears. " Therefore the production of a constant current of electricity is not possible in a closed circuit of bodies belonging to the voltaic series."   Wiedemann further supports this statement by the following theoretical consideration :

" In fact, if a constant electric current were to make its appearance in the circuit, it would produce heat in the metallic conductors themselves, and this heating could at the most be counterbalanced by cooling at the metallic junctions.   In any case it would give rise to an uneven distribution of heat ; moreover an electro-magnetic motor could be driven continuously by the current without any sort of supply from outside, and thus work would be performed, which is impossible, since on firmly joining the metals, for instance by soldering, no further changes to compensate for this work could take place even at the contact surfaces."

And not content with the theoretical and experimental proof that the contact electricity of metals by itself cannot produce any current, we shall see too that Wiedemann finds himself compelled to put forward a special hypothesis to abolish its activity even where it might perhaps make itself evident in the current.

Let us, therefore, try another way of passing from contact electricity to the current. Let us imagine, with Wiedemann, " two metals, such as a zinc rod and a copper rod, soldered together at one end, but with their free ends connected by a third body that does *not* act electromotively in relation to the two metals, but only conducts the opposing electricities collected on its surfaces, so that they are neutralised in it. Then the electric force of separation would always restore the previous difference of potential, thus a constant electric current would make its appearance in the circuit, a current that would be able to perform work without any compensation, which again is impossible.—Accordingly, there cannot be a body which only conducts electricity without electromotive activity in relation to the other bodies." We are no better off than before : the impossibility of creating motion again bars the way. By the contact of chemically indifferent bodies, hence by contact electricity as such, we shall never produce a current.

Let us therefore go back again and try a third way pointed out by Wiedemann :

" Finally, if we immerse a zinc plate and a copper plate in a liquid that contains a so-called *binary* compound,[1] which therefore can be decomposed into two chemically distinct constituents that completely saturate one another, *e.g.* dilute hydrochloric acid (H+Cl), etc., then according to paragraph 27 the zinc becomes negatively charged and the copper

[1] As we should now say, an electrolyte.

positively. On joining the metals, these electricities neutralise one another through the place of contact, through which, therefore, a *current of positive electricity* flows from the copper to the zinc. Moreover, since the electric force of separation making its appearance on the contact of these two metals carries away the positive electricity *in the same direction*, the effects of the electric forces of separation are *not* abolished as in a closed metallic circuit. *Hence there arises a constant current of positive electricity*, flowing in the closed circuit through the copper-zinc junction in the direction of the latter, and through the liquid from the zinc to the copper. We shall return in a moment (paragraph 34, *et seq.*) to the question how far the individual electric forces of separation present in the enclosed circuit *really* participate in the formation of the current.—A combination of conductors providing such a ' galvanic current ' we term a galvanic element, or also a galvanic battery." (I, p. 45.)

Thus the miracle has been accomplished. By the mere electric contact force of separation, which, according to Wiedemann himself, cannot be effective without energy being supplied from outside, a constant current has been produced. And if we were offered nothing more for its explanation than the above passage from Wiedemann, it would indeed be an absolute miracle. What have we learned here about the process ?

1. If zinc and copper are immersed in a liquid containing a so-called *binary* compound, then, according to paragraph 27, the zinc becomes negatively charged and the copper positively charged. But in the whole of paragraph 27 there is no word of any binary compound. It describes only a simple voltaic element of a zinc plate and copper plate, with a piece of cloth moistened by an *acid* liquid interposed between them, and then investigates, without mentioning any chemical processes, the resulting static-electric charges of the two metals.

Hence, the so-called *binary* compound has been smuggled in here by the back-door.

2. What this binary compound is doing here remains completely mysterious. The circumstance that it " *can* be decomposed into two chemical constituents that fully saturate each other " (fully saturate each other after they have been decomposed ? !) could at most teach us something new if it were *actually to decompose.* But we are not told a word about that, hence for the time being we have to assume that it does *not* decompose, *e.g.* in the case of paraffin.

3. When the zinc in the liquid has been negatively charged, and the copper positively charged, we bring them into contact (outside the liquid). At once " these electricities neutralise one another through the place of contact, through which therefore a current of *positive electricity* flows from the copper to the zinc." Again, we do not learn why only a current of " positive " electricity flows in the one direction, and not also a current of " negative " electricity in the opposite direction. We do not learn at all what becomes of the negative electricity, which, hitherto, was just as necessary as the positive ; the effect of the electric force of separation consisted precisely in setting them free to oppose one another. Now it has been suddenly suppressed, as it were eliminated, and it is made to appear as if there exists only positive electricity.

But then again, on p. 51, the precise opposite is said, for here " *the electricities unite* in one current " ; consequently both negative and positive flow in it ! Who will rescue us from this confusion ?

4. " Moreover, since the electric force of separation making its appearance on the contact with these two metals *carries away* the positive electricity *in the same direction*, the effects of the electric forces of separation are not abolished as in a closed metallic circuit. *Hence,*

there arises a constant current," etc.—This is a bit thick. For as we shall see a few pages later (p. 52), Wiedemann proves to us that on the "formation of a constant current . . . the electric force of separation at the place of contact of the metals . . . *must be inactive*, that not only does a current occur even when this force, instead of carrying away the positive electricity in the same direction, acts in opposition to the direction of the current, but that in this case too it is not compensated by a definite share of the force of separation of the battery and, hence, once again is inactive." Consequently, how can Wiedemann on p. 45 make an electric force of separation participate as a necessary factor in the formation of the current when on p. 52 he puts it out of action for the duration of the current, and that, moreover, by a hypothesis erected specially for this purpose ?

5. " Hence there arises a *constant current* of positive electricity, flowing in the closed circuit from the copper through its place of contact with the zinc, in the direction of the latter, and through the liquid from the zinc to the copper."—But in the case of such a constant electric current, " heat would be produced by it in the conductors themselves," and also it would be possible for " an electro-magnetic motor to be driven by it and thus work performed," which, however, is impossible without supply of energy.   Since Wiedemann up to now has not breathed a syllable as to whether such a supply of energy occurs, or whence it comes, the constant current so far remains just as much an impossibility as in both the previously investigated cases.

No one feels this more than Wiedemann himself. So he finds it desirable to hurry as quickly as possible over the many ticklish points of this remarkable explanation of current formation, and instead to entertain the reader throughout several pages with all kinds of elementary anecdotes about the thermal, chemical, magnetic, and physiological effects of this still mysterious current, in the course of which by way of exception he even adopts a quite popular tone. Then he suddenly continues (p. 49) :

" We have now to investigate in what way the electric forces of separation are active in a closed circuit of two metals and a liquid, *e.g.* zinc, copper, and hydrochloric acid."

" *We know* that when the current traverses the liquid the constituents of the binary compound (HCl) contained in it become separated in such a manner that one constituent (H) *is set free* on the copper, and an equivalent amount of the other (Cl) on the zinc, *whereby* the latter constituent combines with an equivalent amount of zinc to form ZnCl."

We know ! If we know this, we certainly do not know it from Wiedemann who, as we have seen, so far has not breathed a syllable about this process. Further, *if* we do know anything of this process, it is that it cannot proceed in the way described by Wiedemann.

On the formation of a molecule of HCl from hydrogen and chlorine, an amount of energy $=22,000$ units of heat is liberated (Julius Thomsen). Therefore, to break away the chlorine from its combination with hydrogen, the same quantity of energy must be supplied from outside for each molecule of HCl. Where does the battery derive this energy ? Wiedemann's description does not tell us, so let us look for ourselves.

When chlorine combines with zinc to form zinc chloride a considerably greater quantity of energy is liberated than is necessary to separate chlorine from hydrogen ; (Zn,Cl_2) develops 97,210 and $2(H,Cl)$ 44,000 units of heat (Julius Thomsen). With that the process in the battery becomes comprehensible. Hence it is not, as Wiedemann relates, that hydrogen without more ado is liberated from the copper, and chlorine from the zinc, " whereby " then subsequently and accidentally the zinc and chlorine enter into combination. On the contrary, the union of the zinc with the chlorine is the essential, basic condition for the whole process, and as long as this does not take place, one would wait in vain for hydrogen on the copper.

The excess of energy liberated on formation of a molecule of $ZnCl_2$ over that expended on liberating two atoms of H from two molecules of HCl, is converted in the battery into electric motion and provides the entire " electromotive force " that makes its appearance in the current circuit. Hence it is not a mysterious " electric force of separation " that tears asunder hydrogen and chlorine without any demonstrable source of energy, it is the total chemical process taking place in the battery that endows all the " electric forces of separation " and " electromotive forces " of the circuit with the energy necessary for their existence.

For the time being, therefore, we put on record that Wiedemann's *second* explanation of the current gives us just as little assistance as his first one, and let us proceed further with the text :

" This process proves that the behaviour of the binary substance between the metals does not consist merely in a simple predominant attraction of its entire mass for one electricity or the other, as in the case of metals, but that in addition a special action of its constituents is exhibited. Since the constituent Cl is given off where the current of positive electricity enters the fluid, and the constituent H where the negative electricity enters, *we assume* that each equivalent of chlorine in the compound HCl is charged with a definite amount of negative electricity determining its attraction by the entering positive electricity. It is the *electro-negative constituent* of the compound. Similarly the equivalent H must be charged with positive electricity and so represent the electro-positive constituent of the compound. These charges *could* be produced on the combination of H and Cl in just the same way as on the contact of zinc and copper. Since the compound HCl as such is non-electric, *we must assume* accordingly that in it the atoms of the positive and negative constituents contain *equal* quantities of positive and negative electricity.

If now a zinc plate and a copper plate are dipped in dilute hydrochloric acid, *we can suppose* that the zinc has a stronger attraction towards the electro-negative constituent (Cl) than towards the electro-positive one (H). Consequently, the molecules of hydrochloric acid in contact with the zinc *would* dispose themselves so that their electro-negative constituents are turned towards the zinc, and their electro-positive constituents towards the copper. Owing to the constituents when so arranged exerting their electrical attraction on the constituents of the next molecules of HCl, the whole series of molecules between the zinc and copper plates becomes arranged as in Fig. 10 :

—Zinc										Copper +
	−	+	−	+	−	+	−	+	−	+
	Cl	H	Cl	H	Cl	H	Cl	H	Cl	H

If the second metal acts on the positive hydrogen as the zinc does on the negative chlorine, it would help to promote the arrangement. If it acted in the opposite manner, only more weakly, at least the direction would remain unaltered.

By the influence exerted by the negative electricity of the electro-negative constituent Cl adjacent to the zinc, the electricity *would* be so distributed in the zinc that places on it which are close to the Cl of the immediately adjacent atom of acid would become charged positively, those farther away negatively.

Similarly, negative electricity would accumulate in the copper next to the electro-positive constituent (H) of the adjacent atom of hydrochloric acid, and the positive electricity would be driven to the more remote parts.

Next, the positive electricity in the zinc *would* combine with the negative electricity of the immediately adjacent atom of Cl, and the latter itself with the zinc, to form non-electric $ZnCl_2$. The electropositive atom H, which was previously combined with this atom of Cl, *would* unite with the atom of Cl turned towards it belonging to the second atom of HCl, with simultaneous combination of the electricities contained in these atoms ; similarly, the H of the second atom of HCl *would combine* with the Cl of the third atom, and so on, until finally an atom of H *would* be set free on the copper, the positive electricity of which would unite with the distributed negative electricity of the copper, so that it escapes in a non-electrified condition." This process would " repeat itself until the repulsive action of the electricities accumulated in the metal plates on the electricities of the hydrochloric acid constituents turned towards them balances the chemical attraction of the latter by the metals. If, however, the metal plates are joined by a conductor, the free electricities of the metal plates unite with one another and the above-mentioned processes can recommence. *In this way* a constant current of electricity comes into being. —It is evident that in the course of it a continual loss of *vis viva* occurs, owing to the constituents of the binary compound on their migration to the metals moving to the latter with a definite velocity and then coming to rest, either with formation of a compound $(ZnCl_2)$ or by escaping in the free state (H). (*Note* [*by Wiedemann*]: Since the gain in *vis viva* on separation of the constituents Cl and H . . . is compensated by the *vis viva* lost on the union of these constituents with the constituents of the adjacent atoms, the influence of this process can be neglected.) This loss of *vis viva* is equivalent to the quantity of heat which is set free in the visibly occurring chemical process, essentially, therefore, that produced on the solution of an equivalent of zinc in the dilute acid. This value must be the same as that of the work expended on separating the electricities. If, therefore, the electricities unite to form a current, then, during the solution of an equivalent of zinc and the giving off of an equivalent of hydrogen from the liquid, there must make its appearance in the whole circuit, whether in the form of heat or in the form of external performance of work, an amount of work that is likewise equivalent to the development of heat corresponding to this chemical process."

" Let us assume—could—we must assume—we can suppose—would be distributed—would become charged," etc., etc. Sheer conjecture and subjunctives from which only three actual indicatives can be definitely extracted : firstly, that the combination of the zinc with the chlorine is *now* pronounced to be the condition for the liberation of hydrogen ; secondly, as we now learn right at the end and as it were incidentally, that the energy herewith liberated is the source, and indeed the exclusive source, of all energy required for formation of the current ; and thirdly, that this explanation of the current formation is as directly in contradiction to both those previously given as the latter are themselves mutually contradictory.

Further it is said :

" For the formation of a constant current, therefore, there is active wholly and solely the electric force of separation which is derived from the unequal attraction and polarisation of the atoms of the binary compound in the exciting liquid of the battery by the metal electrodes ; at the place of contact of the metals, at which no further mechanical changes can occur, the electric force of separation *must on the other hand be inactive.* That this force, if by chance it *counteracts* the electromotive excitation of the metals by the liquid (as on immersion of zinc and lead in potassium cyanide solution), is not compensated by a definite share of the force of separation at the place of contact, is proved by the above-mentioned complete proportionality of the total electric force of separation (and electromotive force) in the circuit, with the above-mentioned heat equivalent of the chemical process. Hence it must be neutralised in another way. This would most simply occur on the assumption that on contact of the exciting liquid with the metals the electromotive force is produced in a double manner ; on the one hand by an unequally strong attraction of the *mass* of the liquid as a whole towards one or the other electricity, on the other hand by the unequal attraction of the metals towards the *constituents* of the liquid charged with opposite electricities. . . . Owing to the former unequal (mass) attraction towards the electricities, the liquids would fully conform to the law of the voltaic series of metals, and in a closed circuit . . . complete neutralisation to zero of the electric forces of separation (and electromotive forces) take place ; the second (*chemical*) action . . . on the other hand would be provided *solely* by the electric force of separation necessary for formation of the current and the corresponding electromotive force." (I, pp. 52–3.)

Herewith the last relics of the contact theory are now happily eliminated from formation of the current, and simultaneously also the last relics of Wiedemann's first explanation of current formation given on p. 45. It is finally conceded without reservation that the galvanic battery is a simple apparatus for converting liberated chemical energy into electric motion, into so-called electric force of separation and electromotive force, in exactly the same way as the steam engine is an apparatus for converting heat energy into mechanical motion. In the one case, as in the other, the apparatus provides only the conditions for liberation and further transformation of the energy, but supplies no energy on its own account. This once established, it remains for us now to make a closer examination of this third version of Wiedemann's explanation of the current.

How are the energy transformations in the circuit of the battery represented here ?

It is evident, he says, that in the battery

" a continual loss of *vis viva* occurs, owing to the constituents of the binary compound on their migration to the metals moving to the latter with a definite velocity and then coming to rest, either with formation of a compound ($ZnCl_2$) or by escaping in the free state (H).

This loss is equivalent to the quantity of heat which is set free in the visibly occurring chemical process, essentially, therefore, that produced on the solution of an equivalent of zinc in the dilute acid."

Firstly, if the process goes on in *pure* form, no heat at all is set free in the battery on solution of the zinc; the liberated energy is indeed converted directly into electricity and only from this converted once again into heat by the resistance of the whole circuit.

Secondly, *vis viva* is half the product of the mass and the square of the velocity. Hence the above statement would read : the energy set free on solution of an equivalent of zinc in dilute hydrochloric acid, =so many calories, is likewise equivalent to half the product of the mass of the ions and the square of the velocity with which they migrate to the metals. Expressed in this way, the sentence is obviously false ; the *vis viva* appearing on the migration of the ions is far removed from being equivalent to the energy set free by the chemical process.[1] But if it were to be so, no current would be possible, since there would be no energy remaining over for the current in the remainder of the circuit. Hence the further remark is introduced that the ions come to rest " either with formation of a compound (ZnCl₂) or by escaping in the free state." But if the loss of *vis viva* is to include also the energy changes taking place on these two processes, then we have indeed arrived at a deadlock. For it is precisely to these two processes taken together that we owe the whole liberated energy, so that there can be absolutely no question here of a *loss* of *vis viva*, but at most of a *gain*.

It is therefore obvious that Wiedemann himself did not mean anything definite by this sentence, rather the " loss of *vis viva* " represents only the *deus ex machina* which is to enable him to make the fatal leap from the old contact theory to the chemical explanation of the current. In point of fact, the loss of *vis viva* has now performed its function and is dismissed; henceforth the chemical process in the battery has undisputed sway

---

[1] F. Kohlrausch has recently calculated (Wiedemann's *Annalen*, VI, p. 206) that " immense forces " are required to drive the ions through the water solvent. To cause one milligram to move through a distance of one millimetre requires an attractive force which for H = 32,500 kg., for Cl = 5,200 kg., hence for HCl = 37,700 kg.—Even if these figures are absolutely correct, they do not affect what has been said above. But the calculation contains the hypothetical factors hitherto inevitable in the sphere of electricity and therefore requires control by experiment.* Such control appears possible. In the first place, these " immense forces " must reappear as a definite quantity of energy in the place where they are consumed, *i.e.* in the above case in the water. Secondly, the energy consumed by them must be smaller than that supplied by the chemical processes of the battery, and there should be a definite difference. Thirdly, this difference must be used up in the rest of the circuit and likewise be quantitatively demonstrable there. Only after confirmation by this control can the above figures be regarded as final. The demonstration in the decomposition cell appears still more susceptible of realisation. (*Note by F. Engels.*)

* Actually the hypothesis was incorrect. It is now believed that when HCl is dissolved in water, it is almost completely broken up into positive H ions and negative Cl ions, which do not require " immense forces " to drive them. Engels was fully justified in his scepticism.

as the sole source of energy for current formation, and the only remaining anxiety of our author is as to how he can politely get rid from the current of the last relic of excitation of electricity by the contact of chemically indifferent bodies, namely, the force of separation active at the place of contact of the two metals.

Reading the above explanation of current formation given by Wiedemann, one could believe oneself in the presence of a specimen of the kind of apologia that wholly- and half-credulous theologians of almost forty years ago employed to meet the philologico-historical bible criticism of Strauss, Wilke, Bruno Bauer, etc. The method is exactly the same, and it is bound to be so. For in both cases it is a question of saving the *heritage of tradition* from scientific thought. Exclusive empiricism, which at most allows thinking in the form of mathematical calculation, imagines that it operates only with undeniable facts. In reality, however, it operates predominantly with out-of-date notions, with the largely obsolete products of thought of its predecessors, and such are positive and negative electricity, the electric force of separation, the contact theory. These serve it as the foundation of endless mathematical calculations in which, owing to the strictness of the mathematical formulation, the hypothetical nature of the premises gets comfortably forgotten. This kind of empiricism is as credulous towards the results of the thought of its predecessors as it is sceptical in its attitude to the results of contemporary thought. For it the experimentally established facts have gradually become inseparable from the traditional interpretation associated with them ; the simplest electric phenomenon is presented falsely, *e.g.* by smuggling in the two electricities ; this empiricism *cannot* any longer describe the facts correctly, because the traditional interpretation is woven into the description. In short, we have here in the field of the theory of electricity a tradition just as highly developed as that in the field of theology. And since in both fields the results of recent research, the establishment of hitherto unknown or disputed facts and of the necessarily following theoretical conclusions, run pitilessly counter to the old traditions, the defenders of these traditions find themselves in the direst dilemma. They have to resort to all kinds of subterfuges and untenable expedients, to the glossing over of irreconcilable contradictions, and thus finally land themselves into a medley of contradictions from which they have no escape. It is this faith in all the old theory of electricity that entangles Wiedemann here in the most hopeless contradictions, simply owing to the hopeless attempt to reconcile rationally the old explanation of the current by " contact force," with the modern one by liberation of chemical energy.

It will perhaps be objected that the above criticism of Wiedemann's explanation of the current rests on juggling with words. It may be objected that, although at the beginning Wiedemann expresses himself somewhat carelessly and inaccurately, still he does finally give the correct account in accord with the principle of the conservation of energy and so sets everything right. As against this view, we give below another example, his description of the process in the battery : zinc—dilute sulphuric acid—copper :

" If, however, the two plates are joined by a wire, a galvanic current arises. . . . *By the electrolytic process*, one equivalent of hydrogen is given off at the copper plate from the water of the dilute sulphuric acid, this hydrogen escaping in bubbles. At the zinc there is formed one equivalent of oxygen which oxidises the zinc to form zinc oxide, the latter becoming dissolved in the surrounding acid to form sulphuric zinc oxide." (I, pp. 592–3.)

To break up water into hydrogen and oxygen requires an amount of energy of 69,924 heat-units for each molecule of water. From where then comes the energy in the above cell ? " By the electrolytic process." And from where does the electrolytic process get it ? No answer is given.

But Wiedemann further tells us, not once, but at least twice (I, p. 472 and p. 614), that " according to recent knowledge the water itself is not decomposed," but that in our case it is the sulphuric acid H_2SO_4 that splits up into H_2 on the one hand and into SO_3+O on the other hand, whereby under suitable conditions H_2 and O can escape in gaseous form. But this alters the whole nature of the process. The H_2 of the H_2SO_4 is directly replaced by the bivalent zinc, forming zinc sulphate, $ZnSO_4$. There remains over, on the one side H_2, on the other SO_3+O. The two gases escape in the proportions in which they unite to form water, the SO_3 unites with the water of the solvent to reform H_2SO_4, i.e. sulphuric acid. The formation of $ZnSO_4$, however, develops sufficient energy not only to replace and liberate the hydrogen of the sulphuric acid, but also to leave over a considerable excess, which in our case is expended in forming the current. Hence the zinc does not wait until the electrolytic process puts free oxygen at its disposal, in order first to become oxidised and then to become dissolved in the acid. On the contrary, it enters directly into the process, which only comes into being at all *by this participation of the zinc.*

We see here how obsolete chemical notions come to the aid of the obsolete contact notions. According to modern views, a salt is an acid in which hydrogen has been replaced by a metal. The process under investigation confirms this view , the direct replacement of the hydrogen of the acid by the zinc fully explains the energy change. The old view, adhered to by Wiedemann, regards a salt as a compound of a metallic oxide with an acid and therefore speaks of sulphuric zinc oxide instead of zinc sulphate. But to arrive at sulphuric zinc oxide in our battery of zinc and sulphuric acid, the zinc must first be oxidised. In order to oxidise the zinc fast enough, we must have free oxygen. In order to get free oxygen, we must assume—since hydrogen appears at the copper plate—that the water is decomposed. In order to decompose water, we need tremendous energy. How are we to get this ? Simply " by the electrolytic process " which itself cannot come into operation as long as its chemical end product, the " sulphuric zinc oxide," has not begun to be formed. The child gives birth to the mother.

Consequently, here again Wiedemann puts the whole course of the process absolutely the wrong way round and upside down. And the reason is that he lumps together active and passive electrolysis, two directly opposite processes, simply as electrolysis.

So far we have only examined the events in the battery, i.e. that process in which an excess of energy is set free by chemical action and is converted into electricity by the arrangements of the battery. But it is well known that this process can also be reversed : the electricity of a constant current produced in the battery from chemical energy can, in its turn, be reconverted into chemical energy in a decomposition cell inserted in the circuit. The two processes are obviously the opposites of each other ; if the first is regarded as chemico-electric, then the second is electro-chemical. Both can take place in the same circuit with the same substances. Thus, the voltaic pile from gas elements, the current of which is produced by the union of hydrogen and oxygen to form water, can, in a decomposition cell inserted in the circuit, furnish hydrogen and oxygen in the proportion in which they form water. The usual mode of view lumps these two opposite processes together under the single expression : electrolysis, and does not even distinguish between active and passive electrolysis, between an exciting liquid and a passive electrolyte. Thus Wiedemann treats of electrolysis in general for 143 pages and then adds at the end some remarks on " electrolysis in the battery," in which, moreover, the processes in actual batteries only occupy the lesser part of the seventeen pages of this section. Also in the " theory of electrolysis " that follows, this contrast of battery and decomposition cell is not even mentioned, and anyone who looked for some treatment of the energy changes in the circuit in the next chapter, " the influence of electrolysis on the conduction resistance and the electromotive force in the circuit " would be bitterly disappointed.

Let us now consider the irresistible " electrolytic process " which is able to separate H_2 from O without visible supply of energy, and which plays the same role in the present section of the book as did previously the mysterious " electric force of separation."

" Besides the *primary, purely electrolytic* process of separation of the ions, a quantity of *secondary, purely chemical* processes, quite independent of the first, take place by the action of the ions split off by the current. This action can take place on the material of the electrodes and on the bodies that are decomposed, and in the case of solutions also on the solvent." (I, p. 481.) Let us return to the above-mentioned battery : zinc and copper in dilute sulphuric acid. Here, according to Wiedemann's own statement, the separated ions are the H_2 and O of the water. Consequently for him the oxidation of the zinc and the formation of $ZnSO_4$ is a secondary, purely chemical process, independent of the electrolytic process, in spite of the fact that it is only through it that the primary process becomes possible.

Let us now examine somewhat in detail the confusion that must necessarily arise from this inversion of the true course of events.

Let us consider in the first place the so-called secondary processes in the decomposition cell, of which Wiedemann puts forward some examples [1] (pp. 481, 482).

I. "The electrolysis of Na_2SO_4 dissolved in water. This "breaks up . . . into 1 equivalent of SO_3+O . . . and 1 equivalent of Na. . . . The latter, however, reacts on the water solvent and splits off from it 1 equivalent of H, while 1 equivalent of sodium is formed and becomes dissolved in the surrounding water."

The equation is

$$Na_2SO_4+2H_2O=O+SO_3+2NaOH+2H.$$

In fact, in this example the decomposition
$$Na_2SO_4=Na_2+SO_3+O$$

could be regarded as the primary electro-chemical process, and the further transformation

$$Na_2+2H_2O=2NaHO+2H$$

as the secondary, purely chemical one. But this secondary process is effected immediately at the electrode where the hydrogen appears, the very considerable quantity of energy (111,810 heat-units for Na, O, H, aq. according to Julius Thomsen) thereby liberated is therefore, at least for the most part, converted into electricity, and only a portion in the cell is transformed directly into heat. But the latter can also happen to the chemical energy directly or primarily liberated in the *battery*. The quantity of energy which has thus

[1] It may be noted here once for all that Wiedemann employs throughout the old chemical equivalent values, writing HO, ZnCl, etc. In my equations, the modern atomic weights are everywhere employed, putting, therefore, H_2O, $ZnCl_2$, etc. [*Note by F. Engels.*]

become available and converted into electricity, however, is to be subtracted from that which the current
has to supply for continued decomposition of the
Na_2SO_4. If the conversion of sodium into hydrated
oxide appeared in the *first* moment of the total process
as a secondary process, from the second moment onwards it becomes an essential factor of the total process
and so ceases to be secondary.

But yet a third process takes place in this decomposition cell : SO_3 combines with H_2O to form H_2SO_4,
sulphuric acid, provided the SO_3 does not enter into
combination with the metal of the positive electrode, in
which case again energy would be liberated. But this
change does not necessarily proceed immediately at
the electrode, and consequently the quantity of energy
(21,320 heat-units, J. Thomsen) thereby liberated
becomes converted wholly or mainly into heat in the
cell itself, and provides at most a very small portion
of the electricity in the current. The only really
secondary process occurring in this cell is therefore not
mentioned at all by Wiedemann.

II. " If a solution of copper sulphate is electrolysed
between a positive copper electrode and a negative
one of platinum, 1 equivalent of copper separates out
for 1 equivalent of water decomposed at the negative
platinum electrode, with simultaneous decomposition
of sulphuric acid in the same current circuit ; at the
positive electrode, 1 equivalent of SO_4 should make its
appearance ; but this combines with the copper of the
electrode to form one equivalent of $CuSO_4$, which
becomes dissolved in the water of the electrolysed
solution."

In the modern chemical mode of expression we have,
therefore, to represent the process as follows : copper is
deposited on the platinum ; the liberated SO_4, which
cannot exist by itself, splits up into SO_3+O, the latter

escaping in the free state ; the SO_3 takes up H_2O from the aqueous solvent and forms H_2SO_4, which again combines with the copper of the electrode to form $CuSO_4$, H_2 being set free. Accurately speaking, we have here three processes : (1) the separation of Cu and SO_4 ; (2) $SO_3 + O + H_2O = H_2SO_4 + O$; (3) $H_2SO_4 + Cu = H_2 + CuSO_4$. It is natural to regard the first as primary, the two others as secondary. But if we inquire into the energy changes, we find that the first process is completely compensated by a part of the third : the separation of copper from SO_4 by the reuniting of both at the other electrode. If we leave out of account the energy required for shifting the copper from one electrode to the other, and likewise the inevitable, not accurately determinable, loss of energy in the cell by conversion into heat, we have here a case where the so-called primary process withdraws no energy from the current. The current provides energy exclusively to make possible the separation of H_2 and O, which moreover is indirect, and this proves to be the real chemical result of the whole process—hence, for carrying out a *secondary*, or even tertiary, process.

Nevertheless, in both the above examples, as in other cases also, it is undeniable that the distinction of primary and secondary processes has a relative justification. Thus in both cases, among other things, water also is apparently decomposed and the elements of water given off at the opposite electrodes. Since, according to the most recent experiments, absolutely pure water comes as near as possible to being an ideal non-conductor, hence also a non-electrolyte, it is important to show that in these and similar cases it is not the water that is directly electro-chemically decomposed, but that the elements of water are separated from the acid, in the formation of which here it is true the water solvent must participate.

III. " If one electrolyses hydrochloric acid simultaneously in two U-tubes . . . using in one tube a
zinc positive electrode and in the other tube one of
copper, then in the first tube a quantity of zinc
32·53 is dissolved, in the other a quantity of copper
$2 \times 32·7$."

For the time being let us leave the copper out of
account and consider the zinc. The decomposition of
HCl is regarded here as the primary process, the solution
of Zn as secondary.

According to this conception, therefore, the current
brings to the decomposition cell from outside the energy
necessary for the separation of H and Cl, and after this
separation is completed the Cl combines with the Zn,
whereby a quantity of energy is set free that is subtracted
from that required for separating H and Cl; the current
needs only therefore to supply the difference. So far
everything agrees beautifully ; but if we consider the
two amounts of energy more closely we find that the
one liberated on the formation of $ZnCl_2$ is *larger* than
that used up in separating 2HCl ; consequently, that
the current not only does not need to supply energy,
but on the contrary *receives energy*. We are no longer
confronted by a passive electrolyte, but by an exciting
fluid, not a decomposition cell but a *battery*, which
strengthens the current-forming voltaic pile by a new
element ; the process which we are supposed to conceive
as secondary becomes absolutely primary, becoming the
source of energy of the whole process and making the
latter independent of the current supplied by the voltaic
pile.

We see clearly here the source of the whole confusion
prevailing in Wiedemann's theoretical description.
Wiedemann's point of departure is electrolysis , whether
this is active or passive, battery or decomposition cell, is
all one to him : saw-bones is saw-bones, as the sergeant-

major said to the doctor of philosophy doing his year's military service. And since it is easier to study electrolysis in the decomposition cell than in the battery, he does, in fact, take the decomposition cell as his point of departure, and he makes the processes taking place in it, and the partly justifiable division of them into primary and secondary, the measure of the altogether reverse processes in the battery, not even noticing when his decomposition cell becomes surreptitiously transformed into a battery. Hence he is able to put forward the statement : "the chemical affinity that the separated substances have for the electrodes has no influence on the electrolytic process as such" (I, p. 471), a sentence which in this absolute form, as we have seen, is totally false. Hence, further, his threefold theory of current formation : firstly, the old traditional one, by means of pure contact ; secondly, that derived by means of the abstractly conceived electric force of separation, which in an inexplicable manner obtains for itself or for the "electrolytic process" the requisite energy for splitting apart the H and Cl in the battery and for forming a current as well ; and finally, the modern, chemico-electric theory which demonstrates the source of this energy in the algebraic sum of the chemical reactions in the battery. Just as he does not notice that the second explanation overthrows the first, so also he has no idea that the third in its turn overthrows the second. On the contrary, the principle of the conservation of energy is merely added in a quite superficial way to the old theory handed down from routine, just as a new geometrical theorem is appended to an earlier one. He has no inkling that this principle makes necessary a revision of the whole traditional point of view in this as in all other fields of natural science. Hence Wiedemann confines himself to noting the principle in his explanation of the current, and then calmly puts it on one side, taking it up again only right at the end of the book, in the chapter on the work performed by the current. Even in the theory of the excitation of electricity by contact (I, p. 781 *et seq.*) the conservation of energy plays no role at all in relation to the chief subject dealt with, and is only incidentally brought in for throwing light on subsidiary matters : it is and remains a " secondary process."

Let us return to the above example III. There the same current was used to electrolyse hydrochloric acid in two U-tubes, but in one there was a positive electrode of zinc, in the other, the positive electrode used was of copper. According to Faraday's basic law of electrolysis, the same galvanic current decomposes in each cell equivalent quantities of electrolyte, and the quantities of the substances liberated at the two electrodes are also in proportion to their equivalents (I, p. 470). In the above case it was found that in the first tube a quantity of zinc $32 \cdot 53$ was dissolved, and in the other a quantity of copper $2 \times 31 \cdot 7$. " Nevertheless," continues Wiedemann, " this is no proof for the equivalence of these values. They are observed only in the case of very weak currents with the formation of zinc chloride . . . on the one hand, and of copper chloride . . . on the other. In the case of denser currents, with the same amount of zinc dissolved, the quantity of dissolved copper would sink with formation of increasing quantities of chloride . . . up to $31 \cdot 7$."

It is well known that zinc forms only a single compound with chlorine, zinc chloride, $ZnCl$; copper on the other hand forms two compounds, cupric chloride, $CuCl_2$, and cuprous chloride, Cu_2Cl_2. Hence the process is that the weak current splits off two copper atoms from the electrode for each two chlorine atoms, the two copper atoms remaining united by *one* of their two valencies, while their two free valencies unite with the two chlorine atoms :

$$Cu—Cl$$
$$|$$
$$Cu—Cl$$

On the other hand, if the current becomes stronger, it splits the copper atoms apart altogether, and each one unites with two chlorine atoms.

In the case of currents of medium strength, both compounds are formed side by side. Thus it is solely the strength of the current that determines the formation of one or the other compound, and therefore the process is essentially *electro*-chemical, if this word has any meaning at all. Nevertheless Wiedemann declares explicitly that it is secondary, hence not electro-chemical, but purely chemical.

The above experiment is one performed by Renault (1867) and is one of a whole series of similar experiments in which the same current is led in one U-tube through salt solution (positive electrode—zinc), and in another cell through a varying electrolyte with various metals as the positive electrode. The amounts of the other metals dissolved here for each equivalent of zinc diverged very considerably, and Wiedemann gives the results of the whole series of experiments which, however, in point of fact, are mostly self-evident chemically and could not be otherwise. Thus, for one equivalent of zinc, only two-thirds of an equivalent of gold is dissolved in hydrochloric acid. This can only appear remarkable if, like Wiedemann, one adheres to the old equivalent weights and writes ZnCl for zinc chloride, according to which both the chlorine and the zinc appear in the chloride with only a *single* valency. In reality two chlorine atoms are included to one zinc atom, $ZnCl_2$, and as soon as we know this formula we see at once that in the above determination of equivalents, the chlorine atom is to be taken as the unit and not the zinc atom. The formula for gold chloride, however, is $AuCl_3$, from which it is at once seen that $3ZnCl_2$ contains exactly as much chlorine as $2AuCl_3$, and so all primary, secondary, and tertiary processes in the battery or cell are compelled to transform, for each part by weight [1] of zinc converted into zinc chloride, neither more nor less than two-thirds of a part by weight of gold into gold chloride. This holds absolutely unless the compound $AuCl_3$ [2] also could be prepared by galvanic means, in which case two equivalents of gold even would have to be dissolved for one equivalent of zinc, when also similar variations according to the current strength could occur as in the case of copper and chlorine mentioned above. The value of Renault's researches consists in the fact that they show how Faraday's law is confirmed by facts that appear to contradict it. But what they are supposed to contribute in throwing light on secondary processes in electrolysis is not evident.

Wiedemann's third example led us again from the decomposition cell to the battery, and in fact the battery offers by far the greatest interest when one investigates the electrolytic processes in relation to the transformations of energy taking place. Thus we not infrequently encounter batteries in which the chemico-electric processes seem to take place in direct contradiction to the law of the conservation of energy and in opposition to chemical affinity.

According to Poggendorff's measurements, the battery : zinc—concentrated salt solution—platinum,

[1] As it stands this is untrue. Probably " part by weight " is a slip of Engels' pen for " equivalent by weight " or some such phrase.

[2] Again this does not make sense as it stands. Presumably Engels meant to refer to a hypothetical AuCl.

provides a current of strength 134·6. Hence we have here quite a respectable quantity of electricity, one-third more than in the Daniell cell. What is the source of the energy appearing here as electricity ? The "primary" process is the replacement of sodium in the chlorine compound by zinc. But in ordinary chemistry it is not zinc that replaces sodium, but *vice versa*, sodium replacing zinc from chlorine and other compounds. The "primary" process, far from being able to give the current the above quantity of energy, on the contrary requires itself a supply of energy from outside in order to come into being. Hence, with the mere "primary" process we are again at a standstill. Let us look, therefore, at the real process. We find that the change is not

$$Zn + 2NaCl = ZnCl_2 + 2Na,$$

but

$$Zn + 2NaCl + 2H_2O = ZnCl_2 + 2NaOH + H_2.$$

In other words, the sodium is not split off in the free state at the negative electrode, but forms a hydroxide as in the above example I (pp. 118–119). To calculate the energy changes taking place here, Julius Thomsen's determinations provide us at least with certain important data. According to them, the energy liberated on combination is as follows :

$(ZnCl_2) = 97,210$, $(ZnCl_2,$ aqua$) = 15,630$, making a total for dissolved

zinc chloride	= 112,840 heat-units.
2 (Na, O, H, aqua)	= 223,620 ,, ,,
	336,460 ,, ,,

Deducting consumption of energy on the separations :

2(Na,Cl, aq.)	= 193,020 heat-units.
2(H_2,O)	= 136,720 ,, ,,
	329,740 ,, ,,

The excess of liberated energy equals 6,720 heat-units.

This amount is obviously small for the current strength obtained, but it suffices to explain, on the one hand, the separation of the sodium from chlorine, and on the other hand, the current formation in general.

We have here a striking example of the fact that the distinction of primary and secondary processes is purely relative and leads us *ad absurdum* as soon as we take it absolutely. The primary electrolytic process, taken alone, not only cannot produce any current, but cannot even take place itself. It is only the secondary, ostensibly purely chemical process that makes the primary one possible and, moreover, supplies the whole surplus energy for current formation. In reality, therefore, it proves to be the primary process and the other the secondary one. When the rigid differences and opposites, as imagined by the metaphysicians and metaphysical natural scientists, were dialectically reversed into their opposites by Hegel, it was said that he had twisted the words in their mouths. But if nature itself proceeds exactly like old Hegel, it is surely time to examine the matter more closely.

With greater justification one can regard as secondary those processes which, while taking place *in consequence* of the chemico-electric process of the battery or the electrochemical process of the decomposition cell, do so independently and separately, occurring therefore at the same distance from the electrodes. The energy changes taking place in such secondary processes likewise do not enter into the electric process; directly they neither withdraw energy from it nor supply energy to it. Such processes occur very frequently in the decomposition cell; we saw an instance in the example I above on the formation of sulphuric acid during electrolysis of sodium sulphate. They are, however, of lesser interest here. Their occurrence in the battery, on the other hand, is of greater practical importance. For although they do not directly supply energy to, or withdraw it from, the chemico-electric process, nevertheless they alter the total available energy present in the battery and thus affect it indirectly.

There belong here, besides subsequent chemical changes of the ordinary kind, the phenomena that occur when the ions are liberated at the electrodes in a different condition from that in which they usually occur in the free state, and when they pass over to the latter only after moving away from the electrodes. In such cases the ions can assume a different density or a different state of aggregation. They can also undergo considerable changes in regard to their molecular constitution, and this case is the most interesting. In all these cases, an analogous heat change corresponds to the secondary chemical or physical change of the ions taking place at a certain distance from the electrodes ; usually heat is set free, in some cases it is consumed. This heat change is, of course, restricted in the first place to the place where it occurs : the liquid in the battery or decomposition cell becomes warmer or cooler while the rest of the circuit remains unaffected. Hence this heat is called *local* heat. The liberated chemical energy available for conversion into electricity is, therefore, diminished or increased by the equivalent of this positive or negative local heat produced in the battery. According to Favre, in a battery with hydrogen peroxide and hydrochloric acid two-thirds of the total energy set free is consumed as local heat ; the Grove cell, on the other hand, on closing the circuit became considerably cooler and therefore supplied energy from outside to the circuit by absorption of heat. Hence we see that these secondary processes also react on the primary one. We can make whatever approach we like; the distinction between primary and secondary processes remains merely a relative one and is regularly suspended in the interaction of the one with the other. If this is forgotten and such relative opposites treated as absolute, one finally gets hopelessly involved in contradictions, as we have seen above.

As is well known, on the electrolytic separation of gases the metal electrodes become covered with a thin layer of gas; in consequence the current strength decreases until the electrodes are saturated with gas, whereupon the weakened current again becomes constant. Favre and Silbermann have shown that local heat arises also in such a decomposition cell; this local heat, therefore, can only be due to the fact that the gases are not liberated at the electrodes in the state in which they usually occur, but that they are only brought into their usual state, after their separation from the electrode, by a further process bound up with the development of heat. But what is the state in which the gases are given off at the electrodes ? It is impossible to express oneself more cautiously on this than Wiedemann does. He terms it " a certain," an " allotropic," an " active," and finally, in the case of oxygen, several times an " ozonised " state. In the case of hydrogen his statements are still more mysterious. Incidentally, the view comes out that ozone and hydrogen peroxide are the forms in which this " active " state is realised. Our author is so keen in his pursuit of ozone that he even explains the extreme electro-negative properties of certain peroxides from the fact that they possibly " contain a part of the oxygen in the *ozonised state !* " (I, p. 57.) Certainly both ozone and hydrogen peroxide are formed on the so-called decomposition of water, but only in small quantities. There is no basis at all for assuming that in the case mentioned local heat is produced first of all by the origin and then by the decomposition of any large quantities of the above two compounds. We do not know the heat of formation of ozone, O_3, from *free* oxygen atoms. According to Berthelot the heat of formation of hydrogen peroxide from H_2O (liquid)$+O=-21,480$; the origin of this compound in any large amount would therefore give rise to a large excess of energy (about 30 per cent. of the energy required for the separation of H_2 and O), which could not but be evident and demonstrable. Finally, ozone and hydrogen peroxide would only take oxygen into account (apart from current reversals, where both gases would come together at the same electrode), but not hydrogen. Yet the latter also escapes in an " active " state, so much so that in the combination : potassium nitrate solution between platinum electrodes, it combines directly with the nitrogen split off from the acid to form ammonia.

In point of fact, all these difficulties and doubts have no existence. The electrolytic process has no monopoly of splitting off bodies " in an active state." Every chemical decomposition does the same thing. It splits off the liberated chemical elements in the first place in the form of free atoms of O, H, N, etc., which only after their liberation can unite to form molecules, O_2, H_2, N_2, etc., and on thus uniting give off a definite, though up-to-now still undetermined,[1] quantity of energy which appears as heat. But during the infinitesimal moment of time when the atoms are free, they are the bearers of the total quantity of energy that they can take up at all ; while possessed of their maximum energy they are free to enter into any combination offered them. Hence they are " in an active state " in contrast to the molecules O_2, H_2, N_2, which have already surrendered a part of this energy and cannot enter into combination with other elements without this quantity of energy surren-

---

[1] This quantity has now not only been determined but utilised. Thus if the hydrogen is previously split into atoms, the ordinary oxy-hydrogen flame can be made a great deal hotter.

dered being re-supplied from outside. We have no need, therefore, to resort to ozone and hydrogen peroxide, which themselves are only products of this active state. For instance, we can undertake the abovementioned formation of ammonia on electrolysis of potassium nitrate even without a battery, simply by chemical means, by adding to nitric acid or a nitrate solution a liquid in which hydrogen is set free by a chemical process. In both cases the active state of the hydrogen is the same. But the interesting point about the electrolytic process is that here the transitory existence of the free atoms becomes as it were tangible. The process here is divided into two phases : the electrolysis provides free atoms at the electrodes, but their combination to form molecules occurs at some distance from the electrodes. However infinitesimally minute this distance may be compared to measurements where masses are concerned, it suffices to prevent the energy liberated on formation of the molecules being used for the electric process, at least for the most part, and so determines its conversion into heat—the local heat in the battery. But it is owing to this that the fact is established that the elements are split off as free atoms and for a moment have existed in the battery as free atoms. This fact, which in pure chemistry can only be established by theoretical conclusions,[1] is here proved experimentally, in so far as this is possible without sensuous perception of the atoms and molecules themselves. Herein lies the high scientific importance of the so-called local heat of the battery.

The conversion of chemical energy into electricity by means of the battery is a process about whose course we know next to nothing, and which we shall get to know in more detail only when the *modus operandi* of electric motion itself becomes better known.

[1] It has since been proved experimentally.

The battery has ascribed to it an " electric force of separation " which is given for each particular battery. As we saw at the outset, Wiedemann conceded that this electric force of separation is not a definite form of energy.   On the contrary, it is primarily nothing more than the capacity, the property, of a battery to convert a definite quantity of liberated chemical energy into electricity in unit time.   Throughout the whole course of events, this chemical energy itself never assumes the form of an " electric force of separation," but, on the contrary, at once and immediately takes on the form of so-called " electromotive force " *i.e.* of electric motion. If in ordinary life we speak of the force of a steam engine in the sense that it is capable in unit time of converting a definite quantity of heat into the motion of masses, this is not a reason for introducing the same confusion of ideas into scientific thought also.   We might just as well speak of the varying force of a pistol, a carbine, a smooth-bored gun, and a blunderbuss, because, with equal gunpowder charges and projectiles of equal weight, they shoot varying distances.   But here the wrongness of the expression is quite obvious.   Everyone knows that it is the ignition of the gunpowder charge that drives the bullet, and that the varying range of the weapon is only determined by the greater or lesser dissipation of energy according to the length of the barrel, the form of the projectile, and the tightness of its fitting. But it is the same for steam power and for the electric force of separation.   Two steam engines—other conditions being equal, *i.e.* assuming the quantity of energy liberated in equal periods of time to be equal in both— or two galvanic batteries, of which the same thing holds good, differ as regards performance of work only owing to their greater or lesser dissipation of energy.   And if until now all armies have been able to develop the technique of firearms without the assumption of a special shooting force of weapons, the science of electricity has absolutely no excuse for assuming an " electric force of separation " analogous to this shooting force, a force which embodies absolutely no energy and which therefore of itself cannot perform a millionth of a milli-gram-metre of work.

The same thing holds good for the second form of this " force of separation," the " electric force of contact of metals " mentioned by Helmholtz. It is nothing but the property of metals to convert on their contact the existing energy of another form into electricity. Hence it is likewise a force that does not contain a particle of energy. If we assume with Wiedemann that the source of energy of contact electricity lies in the *vis viva* of the motion of adhesion, then this energy exists in the first place in the form of this mass motion and on its vanishing becomes converted immediately into electric motion, without even for a moment assuming the form of an " electric force of contact."

And now we are assured in addition that the electro-motive force, *i.e.* the chemical energy, reappearing as electric motion is proportional to this " electric force of separation," which not only contains no energy, but owing to the very conception of it *cannot* contain any ! This proportionality between non-energy and energy obviously belongs to the same mathematics as that in which there figures the " ratio of the unit of electricity to the milligram." But the absurd form, which owes its existence only to the conception of a *simple property* as a *mystical* force, conceals a quite simple tautology : the capacity of a given battery to convert liberated chemical energy into electricity is measured—by what ? By the quantity of the energy reappearing in the circuit as electricity in relation to the chemical energy consumed in the battery. That is all.

In order to arrive at an electric force of separation,

one must take seriously the device of the two electric fluids. To convert this from its neutrality to its polarity, hence to split it apart, requires a certain expenditure of energy—the electric force of separation. Once separated, the two electricities can, on being reunited, again give off the same quantity of energy—electromotive force. But since nowadays no one, not even Wiedemann, regards the two electricities as having a real existence, it means that one is writing for a defunct public if one deals at length with such a point of view.

The basic error of the contact theory consists in the fact that it cannot divorce itself from the idea that contact force or electric force of separation is a *source of energy*, which of course was difficult when the mere capacity of an apparatus to bring about transformation of energy had been converted into a *force*; for indeed, a *force* ought precisely to be a definite form of energy. Because Wiedemann cannot rid himself of this unclear notion of force, although alongside of it the modern ideas of indestructible and uncreatable energy have been forced upon him, he falls into his nonsensical explanation of the current, No. 1, and into all the later demonstrated contradictions.

If the expression " electric force of separation " is directly contrary to reason, the other " electromotive force " is at least superfluous. We had heat engines long before we had electro-motors, and yet the theory of heat has been developed quite well without any special thermo-motor force. Just as the simple expression heat includes all phenomena of motion that belong to this form of energy, so also can the expression electricity in its own sphere. Moreover, very many forms of action of electricity are not at all directly " motor ": the magnetisation of iron, chemical decomposition, conversion into heat. And finally, in every natural science, even in mechanics, it is always an advance if the word *force* can somehow be got rid of.[1]

We saw that Wiedemann did not accept the chemical explanation of the processes in the battery without a certain reluctance. This reluctance continually attacks him ; where he can blame anything on the so-called chemical theory, this is certain to occur. Thus, " it is by no means established that the electromotive force is proportional to the intensity of chemical action." (I, p. 791.) Certainly not in every case ; but where this proportionality does not occur, it is only a proof that the battery has been badly constructed, that dissipation of energy takes place in it. For that reason Wiedemann is quite right in paying no attention in his theoretical deductions to such subsidiary circumstances which falsify the purity of the process, but in simply assuring us that the electromotive force of a cell is equal to the mechanical equivalent of the chemical action taking place in it in unit time with unit intensity of current.

In another passage we read :

" That further, in the acid-alkali battery, the combination of acid and alkali is not the cause of current formation follows from the experiments paragraph 61 (Becquerel and Fechner), paragraph 260 (Dubois-Reymond), and paragraph 261 (Worm-Müller), according to which in certain cases when these are present in equivalent quantities no current makes its appearance, and likewise from the experiments (Henrici) mentioned in paragraph 62, that on inter-posing a solution of potassium nitrate between the potassium hydroxide and nitric acid, the electromotive force makes its appearance in the same way as without this interposition." (I, p. 791.)

[1] This statement has been very fully confirmed by the progress of physics in the last fifty years. It is interesting to note that idealistic writers have used this disappearance of the notion of force as an argument that materialism is being refuted !

The question whether the combination of acid and alkali is the cause of current formation is a matter of very serious concern for our author. Put in this form it is very easy to answer. The combination of acid and alkali is first of all the cause of a *salt* being formed with liberation of energy. Whether this energy wholly or partly takes the form of electricity depends on the circumstances under which it is liberated. For instance, in the battery : nitric acid and potassium hydroxide between platinum electrodes, this will be at least partially the case, and it is a matter of indifference for the *formation* of the current whether a potassium nitrate solution is interposed between the acid and alkali or not, since this can at most delay the salt formation but not prevent it. If, however, a battery is formed like one of Worm-Müller's, to which Wiedemann constantly refers, where the acid and alkali solution is in the middle, but a solution of their salt at both ends, and in the same concentration as the solution that is formed in the battery, then it is obvious that no current can arise, because on account of the end members—since everywhere identical bodies are formed—*no ions can be produced.* Hence the conversion of the liberated energy into electricity has been prevented in as direct a manner as if the circuit had not been closed ; it is therefore not to be wondered at that no current is obtained. But that acid and alkali can in general produce a current is proved by the battery : carbon, sulphuric acid (one part in ten of water), potassium hydroxide (one part in ten of water), carbon, which according to Raoult has a current strength of 73.[1] And that, with suitable arrangement of the battery, acid and alkali can provide a current strength corresponding to the large quantity of energy set free on their combination, is seen from the

[1] In all the following data relating to current strength, the Daniell cell is put = 100. [*Note by F. Engels.*]

fact that the most powerful batteries known depend
almost exclusively on the formation of alkali salts, *e.g.*
that of Wheatstone : platinum, platinic chloride,
potassium amalgam—current strength 230 ; lead peroxide, dilute sulphuric acid, potassium amalgam=326 ;
manganese peroxide instead of lead peroxide=280 ;
in each case, if zinc amalgam was employed instead of
potassium amalgam, the current strength fell almost
exactly by 100. Similarly in the battery : manganese
dioxide, potassium permanganate solution, potassium
hydroxide, potassium, Beetz obtained the current
strength 302, and further : platinum, dilute sulphuric
acid, potassium=293·8 ; Joule : platinum, nitric acid,
potassium hydroxide, potassium amalgam=302. The
" cause " of these exceptionally strong current strengths
is certainly the combination of acid and alkali, or alkali
metal, and the large quantity of energy thereby liberated.

A few pages further on it is again stated :

> " It must, however, be carefully borne in mind that
> the equivalent in work of the whole chemical action
> taking place at the place of contact of the hetero-
> geneous bodies is not to be directly regarded as the
> measure of the electromotive force in the circuit.
> When, for instance, in the acid-alkali battery (*iterum
> Crispinus !*) of Becquerel, these two substances
> combine, when carbon is consumed in the battery :
> platinum, molten potassium nitrate, carbon, when the
> zinc is rapidly dissolved in an ordinary cell of copper,
> impure zinc, dilute sulphuric acid, with formation of
> local currents, then a large part of the work produced
> (it should read : energy liberated) in these chemical
> processes . . . is converted into heat and is thus
> lost for the total current circuit." (I, p. 798.)

All these processes are to be referred to loss of energy in
the battery ; they do not affect the fact that the electric
motion arises from transformed chemical energy, but
only affect the quantity of energy transformed.

Electricians have devoted an endless amount of time and trouble to composing the most diverse batteries and measuring their "electromotive force." The experimental material thus accumulated contains very much of value, but certainly still more that is valueless. For instance, what is the scientific value of experiments in which " water " is employed as the electrolyte, when, as has now been proved by F. Kohlrausch, water is the worst conductor and therefore also the worst electrolyte,[1] and where, therefore, it is not the water but its unknown impurities that caused the process ? And yet, for instance, almost half of all Fechner's experiments depend on such employment of water, even his " *experimentum crucis*," by which he sought to establish the contact theory impregnably on the ruins of the chemical theory. As is already evident from this, in almost all such experiments, a few only excepted, the chemical processes in the battery, which however form the source of the so-called electromotive force, remain practically disregarded. There are, however, a number of batteries whose chemical composition does not allow of any certain conclusion being drawn as to the chemical changes proceeding in them when the current circuit is closed. On the contrary, as Wiedemann (I, p. 797) says, it is " not to be denied that we are by no means in all cases able to obtain an insight into the chemical attractions in the battery." Hence, from the ever more important chemical aspect, all such experiments are valueless in so far as they are not repeated with these processes under control.

In these experiments it is indeed only quite by way of

---

[1] A column of the purest water prepared by Kohlrausch 1 mm. in length offered the same resistance as a copper conductor of the same diameter and a length approximately that of the moon's orbit.   Naumann, *Allgemeine Chemie* [*General Chemistry*], p. 729.*   [*Note by F. Engels.*]

* Appendix II, p. 335.

exception that any account is taken of the energy changes taking place in the battery. Many of them were made before the law of the equivalence of motion was recognised in natural science, but as a matter of custom they continue to be dragged from one textbook into another without being controlled or their value summed up. It has been said that electricity has no inertia (which has about as much sense as saying velocity has no specific gravity), but this certainly cannot be said of the *theory* of electricity.

So far, we have regarded the galvanic cell as an arrangement in which, in consequence of the contact relations established, chemical energy is liberated in some way for the time being unknown, and converted into electricity. We have likewise described the decomposition cell as an apparatus in which the reverse process is set up, electric motion being converted into chemical energy and used up as such. In so doing we had to put in the foreground the chemical side of the process that has been so much neglected by electricians, because this was the only way of getting rid of the lumber of notions handed down from the old contact theory and the theory of the two electric fluids. This once accomplished, the question was whether the chemical process in the battery takes place under the same conditions as outside it, or whether special phenomena make their appearance that are dependent on the electric excitation.

In every science, incorrect notions are, in the last resort, apart from errors of observation, incorrect notions of correct facts. The latter remain even when the former are shown to be false. Although we have discarded the old contact theory, the established facts remain, of which they were supposed to be the explanation. Let us consider these and with them the electric aspect proper of the process in the battery.

It is not disputed that on the contact of heterogeneous

bodies, with or without chemical changes, an excitation of electricity occurs which can be demonstrated by means of an electroscope or a galvanometer. As we have already seen at the outset, it is difficult to establish in a particular battery the source of energy of these in themselves extremely minute phenomena of motion ; it suffices that the existence of such an external source is generally conceded.

In 1850–53, Kohlrausch published a series of experiments in which he assembled the separate components of a battery in pairs and tested the static electric stresses produced in each case ; the electromotive force of the cell should then be composed of the algebraic sum of these stresses. Thus, taking the stress of $Zn/Cu=100$, he calculates the relative strengths of the Daniell and Grove cells as follows :

For the Daniell cell :
$Zn/Cu+amalg.Zn/H_2SO_4+Cu/SO_4=100+149-21=228.$

For the Grove cell :
$Zn/Pt+amalg.\ Zn/H_2SO_4+Pt/HNO_3=107+149+149=405,$

which closely agrees with the direct measurement of the current strengths of these cells. These results, however, are by no means certain. In the first place, Wiedemann himself calls attention to the fact that Kohlrausch only gives the final result but " unfortunately no figures for the results of the separate experiments." In the second place, Wiedemann himself repeatedly recognises that all attempts to determine quantitatively the electric excitation on contact of metals, and still more on contact of metal and fluid, are at least very uncertain on account of the numerous unavoidable sources of error. If, nevertheless, he repeatedly uses Kohlrausch's figures in his calculations, we shall do better not to follow him here, the more so

as another means of determination is available which
is not open to these objections.

If the two exciting plates of a battery are immersed
in the liquid and then joined into a circuit by the
terminals of a galvanometer, according to Wiedemann,
" the initial deflection of its magnetic needle, before
chemical changes have altered the strength of the
electric excitation, is a measure of the sum of electromotive forces in the circuit." Batteries of various
strengths, therefore, give initial deflections of various
strengths, and the magnitude of these initial deflections
is proportional to the current strength of the corresponding batteries.

It looks as if we had here tangibly before our eyes the
" electric force of separation," the " contact force,"
which causes motion independently of any chemical
action. And this in fact is the opinion of the whole
contact theory. In reality we are confronted here by a
relation between electric excitation and chemical
action that we have not yet investigated. In order to
pass to this subject, we shall first of all examine rather
more closely the so-called electromotive law ; in so
doing, we shall find that here also the traditional
contact notions not only provide no explanation, but
once again directly bar the way to an explanation.

If in any cell consisting of two metals and a liquid,
e.g. zinc, dilute hydrochloric acid, and copper, one inserts
a third metal such as a platinum plate, without connecting it to the external circuit by a wire, then the
initial deflection of the galvanometer will be exactly
the same as *without* the platinum plate. Consequently
it has no effect on the excitation of electricity. But it
is not permissible to express this so simply in electromotive language. Hence one reads :

" The sum of the electromotive forces of zinc and
platinum and platinum and copper now takes the

place of the electromotive force of zinc and copper in the liquid. Since the path of the electricities is not perceptibly altered by the insertion of the platinum plate, we can conclude from the identity of the galvanometer readings in the two cases, that the electromotive force of zinc and copper in the liquid is equal to that of zinc and platinum plus that of platinum and copper in the same liquid. This would correspond to Volta's theory of the excitation of electricity between the metals as such. The result, which holds good for all liquids and metals, is expressed by saying: On their electromotive excitation by liquids, metals follow the law of the voltaic series. This law is also given the name of the *electromotive law*." (Wiedemann, I, p. 62.)

In saying that in this combination the platinum does not act at all as an exciter of electricity, one expresses what is simply a fact. If one says that it does act as an exciter of electricity, but in two opposite directions with equal strength so that the effect is neutralised, the fact is converted into a hypothesis merely for the sake of doing honour to the "electromotive force." In both cases the platinum plays the role of a fictitious person.

During the first deflection there is still no closed circuit. The acids, being undecomposed,[1] do not conduct; they can only conduct by means of the ions. If the third metal has no influence on the first deflection, this is simply the result of the fact that it is still *isolated*.

How does the third metal behave *after* the establishment of the constant current and during the latter?

In the voltaic series of metals in most liquids, zinc lies after the alkali metals fairly close to the positive end and platinum at the negative end, copper being

---

[1] This statement is in accord with theory fifty years ago, but incorrect.

between the two. Hence, if platinum is put as above between copper and zinc it is negative to them both. If the platinum had any effect at all, the current in the liquid would have to flow to the platinum both from the zinc and from the copper, that is away from both electrodes to the unconnected platinum ; which would be a *contradictio in adjectio*. The basic condition for the action of several different metals in the battery consists precisely in their being connected among themselves externally to the circuit. An unconnected, superfluous metal in the battery acts as a non-conductor ; it can neither form ions nor allow them to pass through, and without ions we know of no conduction in electrolytes. Hence it is not merely a fictitious person, it even stands in the way by forcing the ions to go round it.

The same thing holds good if we connect the zinc and platinum, leaving the copper unconnected in the middle ; here the latter, if it had any effect at all, would produce a current from the zinc to the copper and another from the copper to the platinum ; hence it would have to act as a sort of intermediary electrode and give off hydrogen on the side turned towards the zinc, which again is impossible.

If we discard the traditional electromotive mode of expression the case becomes extremely simple. As we have seen, the galvanic battery is an apparatus in which chemical energy is liberated and transformed into electricity. It consists as a rule of one or more liquids and two metals as electrodes, which must be connected together by a conductor outside the liquids. This completes the apparatus. Anything else that is dipped unconnected into the exciting liquid, whether metal, glass, resin, or whatever you like, cannot participate in the chemico-electric process taking place in the battery, in the formation of the current, so long as the liquid is not chemically altered ; it can at most *hinder* the process.   Whatever the capacity for exciting electricity of a third metal dipped into the liquid may be, or that of one or both electrodes of the battery, it cannot have any effect so long as this metal is not connected to the circuit outside the liquid.

Consequently, not only is Wiedemann's *derivation*, as given above, of the so-called electromotive law false, but the interpretation which he gives to this law is also false.   One can speak neither of a compensating electromotive activity of the unconnected metal, since the sole condition for such activity is cut off from the outset ; nor can the so-called electromotive law be deduced from a fact which lies outside the sphere of this law.

In 1845, old Poggendorff published a series of experiments in which he measured the electromotive force of various batteries, that is to say the quantity of electricity supplied by each of them in unit time.[1]   Of these experiments, the first twenty-seven are of special value, in each of which three given metals were one after another connected in the same exciting liquid to three different batteries, and the latter investigated and compared as regards the quantity of electricity produced.   As a good adherent of the contact theory, Poggendorff also put the third metal unconnected in the battery in each experiment and so had the satisfaction of convincing himself that in all eighty-one batteries this third metal remained a pure inactive element in the combination.   But the significance of these experiments by no means consists in this fact but rather in the confirmation and establishment of the correct meaning of the so-called electromotive law.

Let us consider the above series of batteries in which zinc, copper, and platinum are connected together in pairs in dilute hydrochloric acid.   Here Poggendorff

---

[1] This is, of course, not electromotive force in the modern sense of the term.

found the quantities of electricity produced to be as follows, taking that of a Daniell cell as 100 :

Zinc-copper	.. ..	78·8
Copper-platinum	..	74·3
Total	.. ..	153·1
Zinc-platinum ..	..	153·7

Thus, zinc in direct connection with platinum produced almost exactly the same quantity of electricity as zinc-copper copper-platinum. The same thing occurred in all other batteries, whatever liquids and metals were employed. When, from a series of metals in the same exciting liquid, batteries were formed in such a way that in each case, according to the voltaic series valid for this liquid, the second, third, fourth, etc., one after the other were made to serve as negative electrodes for the preceding one and as positive electrodes for that which followed, then the sum of the quantities of electricity produced by all these batteries is equal to the quantity of electricity produced by a battery formed directly between the two end members of the whole metallic series. For instance, in dilute hydrochloric acid the sum total of the quantities of electricity produced by the batteries zinc-zinc, zinc-iron, iron-copper, copper-silver, and silver-platinum, would be equal to that produced by the battery : zinc-platinum. A pile formed from all the cells of the above series would, other things being equal, be exactly neutralised by the introduction of a zinc-platinum cell with a current of the opposite direction.

In this form, the so-called electromotive law has a real and considerable significance. It reveals a new aspect of the inter-connection between chemical and electrical action. Hitherto, on investigating mainly the *source* of energy of the galvanic current, this source, the chemical change, appeared as the active side of the process; the electricity was produced from it and therefore appeared primarily as passive. Now this is reversed. The electric excitation determined by the constitution of the heterogeneous bodies put into contact in the battery can neither add nor subtract energy from the chemical action (other than by conversion of liberated energy into electricity). It can, however, according as the battery is made up, accelerate or slow down this action.

If the battery, zinc-dilute hydrochloric acid-copper, produced in unit time only half as much electricity for the current as the battery, zinc-dilute hydrochloric acid-platinum, this means in chemical terms that the first battery produces in unit time only half as much zinc chloride and hydrogen as the second. *Hence the chemical action has been doubled, although the purely chemical conditions for this action have remained the same.* The electric excitation has become the regulator of the chemical action; it appears now as the active side, the chemical action being the passive side.

Thus, it becomes comprehensible that a number of processes previously regarded as purely chemical now appear as electro-chemical. Chemically pure zinc is not attacked at all by dilute acid, or only very weakly; ordinary commercial zinc, on the other hand, is rapidly dissolved with formation of a salt and production of hydrogen; it contains an admixture of other metals and carbon, which make their appearance in unequal amounts at various places of the surface. Local currents are formed in the acid between them and the zinc itself, the zinc areas forming the positive electrodes and the other metals the negative electrodes, the hydrogen bubbles being given off on the latter. Likewise the phenomenon that when iron is dipped into a solution of copper sulphate it becomes covered with a

layer of copper is now seen to be an electro-chemical
phenomenon, one determined by the currents which arise
between the heterogeneous areas of the surface of the iron.

In accordance with this we find also that the voltaic
series of metals in liquids corresponds on the whole to
the series in which metals replace one another from their
compounds with halogens and acid radicles. At the
extreme negative end of the voltaic series we regularly
find the metals of the gold group, gold, platinum,
palladium, rhodium, which oxidise with difficulty, are
little or not at all attacked by acids, and which are
easily precipitated from their salts by other metals.
At the extreme positive end are the alkali metals which
exhibit exactly the opposite behaviour: they are
scarcely to be split off from their oxides except with the
greatest expenditure of energy; they occur in nature
almost exclusively in the form of salts, and of all the
metals they have by far the greatest affinity for halogens
and acid radicles. Between these two come the other
metals in somewhat varying sequence, but such that on
the whole electrical and chemical behaviour correspond
to one another. The sequence of the separate members
varies according to the liquids and has hardly been
finally established for any single liquid. It is even
permissible to doubt whether there exists such an
absolute voltaic series of metals for any single liquid.
Given suitable batteries and decomposition cells, two
pieces of the same metal can act as positive and negative
electrodes respectively, hence the same metal can be
both positive and negative towards itself. In thermo-
cells which convert heat into electricity, with large
temperature differences at the two junctions, the direction of the current is reversed; the previously positive
metal becomes negative and *vice versa*. Similarly, there
is no absolute series according to which the metals
replace one another from their chemical compounds

with a particular halogen or acid radicle ; in many cases by supplying energy in the form of heat we are able almost at will to alter and reverse the series valid for ordinary temperatures.

Hence we find here a peculiar interaction between chemical action and electricity. The chemical action in the battery, which provides the electricity with the total energy for current formation, is in many cases first brought into operation, and in all cases quantitatively regulated, by the electric charges developed in the battery. If previously the processes in the battery seemed to be chemico-electric in nature, we see here that they are just as much electro-chemical. From the point of view of formation of the *constant current*, chemical action appears to be the primary thing : from the point of view of *excitation* of current it appears as secondary and accessory. The reciprocal action excludes any absolute primary or absolute secondary ; but it is just as much a double-sided process which from its very nature can be regarded from two different standpoints ; to be understood in its totality it must even be investigated from both standpoints one after the other, before the total result can be arrived at. If, however, we adhere onesidedly to a single standpoint as the absolute one in contrast to the other, or if we arbitrarily jump from one to the other according to the momentary needs of our argument, we shall remain entangled in the onesidedness of metaphysical thinking ; the interconnection escapes us and we become involved in one contradiction after another.

We saw above that, according to Wiedemann, the initial deflection of the galvanometer, immediately after dipping the exciting plates into the liquid of the battery and before chemical changes have altered the strength of the electric excitation, is " a measure of the sum of electromotive forces in the circuit."

So far we have become acquainted with the so-called electromotive force as a form of energy, which in our case was produced in an equivalent amount from chemical energy, and which in the further course of the process became reconverted into equivalent quantities of heat, mass motion, etc. Here we learn all at once that the " sum of the electromotive forces in the circuit " is already in existence *before* this energy has been liberated by chemical changes ; in other words, that the electromotive force is nothing but the capacity of a particular cell to liberate a particular quantity of chemical energy in unit time and to convert it into electric motion. As previously in the case of the electric force of separation, so here also the electromotive force appears as a force which does not contain a single spark of energy. Consequently, Wiedemann understands by "electromotive force " two totally different things : on the one hand, the capacity of a battery to liberate a definite quantity of given chemical energy and to convert it into electric motion, on the other hand, the quantity of electric motion itself that is developed. The fact that the two are proportional, that the one is a measure for the other, does not do away with the distinction between them. The chemical action in the battery, the quantity of electricity developed, and the heat in the circuit derived from it, when no other work is performed, are even more than proportional, they are equivalent ; but that does not infringe the diversity between them. The capacity of a steam engine with a given cylinder bore and piston stroke to produce a given quantity of mechanical motion from the heat supplied is very different from this mechanical motion itself, however proportional to it it may be. And while such a mode of speech was tolerable at a time when natural science had not yet said anything of the conservation of energy, nevertheless it is obvious that since the recognition of

this basic law it is no longer permissible to confuse real
active energy in any form with the capacity of an
apparatus to impart this form to energy which is being
liberated. This confusion is a corollary of the confusion
of force and energy in the case of the electric force of
separation ; these two confusions provide a harmonious
background for Wiedemann's three mutually contradictory explanations of the current, and in the last resort
are the basis in general for all his errors and confusions in
regard to so-called " electromotive force."

Besides the above-considered peculiar interaction
between chemical action and electricity there is also a
second point that they have in common which likewise
indicates a closer kinship between these two forms of
motion. Both can exist only for an *infinitesimal*
period. The chemical process takes place suddenly for
each group of atoms undergoing it. It can be prolonged
only by the presence of new material that continually
renews it. The same thing holds for electric motion.
Hardly has it been produced from some other form
of motion before it is once more converted into a third
form ; only the continual readiness of available energy
can produce the constant current, in which at each
moment new quantities of motion assume the form of
energy and lose it again.

An insight into this close connection of chemical and
electric action and *vice versa* will lead to important results
in both spheres of investigation.[1] Such an insight is
already becoming more and more widespread. Among
chemists, Lothar Meyer, and after him Kekulé, have
plainly stated that a revival of the electro-chemical

---

[1] This has, of course, been very completely verified by the researches of the last fifty years. Electrical theory was revolutionised by
Thomson's study of electrical conduction in gases, which led to his
discovery of electrons. And the whole of chemistry, including the
chemistry of such unions as that between carbon and hydrogen, which
at first sight is quite unconnected with electrical phenomena, has been
restated in terms of electrons.

theory in a rejuvenated form is impending. Among electricians also, as indicated especially by the latest works of F. Kohlrausch, the conviction seems finally to have taken hold that only exact attention to the chemical processes in the battery and decomposition cell can help their science to emerge from the blind alley of old traditions.

And in fact one cannot see how else a firm foundation is to be given to the theory of galvanism and so secondarily to that of magnetism and static electricity, other than by a chemically exact general revision of all traditional uncontrolled experiments made from an obsolete scientific standpoint, with exact attention to establishing the energy changes and preliminary rejection of all traditional theoretical notions about electricity.

VII

DIALECTICS OF NATURE.—NOTES.

(Written 1873–82. They are given here as they appear in the MSS. The Notes on pp. 152–223 are reproduced in the sequence indicated by Engels' own numbering of the manuscript pages. The other Notes were recorded by Engels on loose sheets of paper of various sizes. Since their date of origin could be determined only in a very few cases they have been arranged here according to content : 1. Ideas of polarity ; 2. On dialectical logic and philosophy ; 3. Mathematical ; 4. On Mechanics, Physics, and Chemistry ; 5. A sheet giving the main heads of an arrangement of the subject matter dealt with ; 6. Tidal friction. Each of these sections has been begun on a fresh page. Notes written on a single sheet of paper have been left in the sequence in which they were found. Variant readings and sentences or words crossed out by Engels, have not been given. The only exception is on p. 167, lines 22–24, where the words " Here it becomes a phrase as everywhere where, instead of investigating the unanalysed forms of motion," which were crossed out by Engels, have been restored, as otherwise the rest of the sentence is left incomplete.—*Ed.*)

Büchner.—Rise of the tendency. The passing of German philosophy into materialism—control over science abolished—outbreak of shallow materialist popularisation, in which the materialism had to make up for the lack of science. Its flourishing at the time of the deepest degradation of bourgeois Germany and official German science—1850–60. Vogt, Moleschott, Büchner. Mutual assurance. Revival by Darwinism coming into fashion, the fruits of which were immediately reaped by these gentlemen.

One could let them alone and leave them to their not unpraiseworthy if narrow occupation of teaching atheism, etc., to German philosophy but for : 1, abuse directed against philosophy (passages to be quoted[1]), [2] which in

[1] See Appendix II, p. 335.
[2] Büchner is acquainted with philosophy only as a dogmatist, just as he himself is a dogmatist of the shallowest reflection of the German Enlightenment which missed the intellectual movement of the great French materialists (Hegel on this)—just as Nicolai had that of Voltaire.* Lessing's dead dog Spinoza, *Encyclopædia*, Preface, p. 19.
[*Note by F. Engels.*] * See Appendix II, p. 336.

spite of everything is the glory of Germany, and 2, the presumption of applying the nature theory to society and reforming socialism. Thus they compel us to take note of them.

First of all, what do they achieve in their own sphere ? Quotations.

2. Turning point, pages 170–171.[1] Whence this sudden Hegelianism ? Transition to dialectics. Two philosophical tendencies, the metaphysical with fixed categories, the dialectical (especially Aristotle and Hegel) with fluid categories ; the proofs that these fixed opposites of basis and consequence, cause and effect, identity and difference, appearance and essence are untenable, that analysis shows one pole already present in the other *in nuce*, that at a definite point the one pole becomes transformed into the other, and that all logic develops only from these progressing contradictions.—This mystical in Hegel himself, because the category appears as pre-existing and the dialectics of the real world as its mere reflection. In reality it is the reverse : the dialectics of the brain is only the reflection of the forms of motion of the real world, both of nature and of history. Until the end of the last century, indeed until 1830, natural scientists could manage pretty well with the old metaphysics, because real science did not go beyond mechanics —terrestrial and cosmic. Nevertheless confusion had already been introduced by higher mathematics, which regards the eternal truth of lower mathematics as a superseded point of view, often asserting the contrary, and putting forward propositions which appear sheer nonsense to the lower mathematician. The rigid categories disappeared here ; mathematics arrived at a field where even such simple relations as those of mere abstract quantity, bad infinity, assumed a completely dialectical form and compelled the mathematicians to

[1] See Appendix II, p. 336.

become dialectical, unconsciously and against their will. There is nothing more comical than the twistings, subterfuges, and expedients employed by the mathematicians to solve this contradiction, to reconcile higher and lower mathematics, to make clear to their understanding that what they had arrived at as an undeniable result is not sheer nonsense, and in general rationally to explain the starting point, method, and result of the mathematics of the infinite.

Now, however, everything is quite different. Chemistry, the abstract divisibility of physical things, bad infinity—atomistics. Physiology—the cell (the organic process of development, both of the individual and of species, by differentiation, the most striking test of rational dialectics), and finally the identity of the forces of nature and their mutual convertability, which put an end to all fixity of categories. Nevertheless, the bulk of natural scientists are still held fast in the old metaphysical categories and helpless when these modern facts, which so to say prove the dialectics in nature, have to be rationally explained and brought into relation with one another. And here *thinking* is necessary : atoms and molecules, etc., cannot be observed under the microscope, but only by the process of thought. Compare the chemists (except for Schorlemmer, who is acquainted with Hegel) and Virchow's cellular pathology, where in the end the helplessness has to be concealed by general phrases. Dialectics divested of mysticism becomes an absolute necessity for natural science, which has forsaken the field where rigid categories sufficed, as it were the lower mathematics of logic, its everyday weapons. Philosophy takes its revenge posthumously on natural science for the latter having deserted it ; and yet the scientists could have seen even from the successes in natural science achieved by philosophy that the latter possessed something that was superior to them even in their own special sphere. (Leibniz, the founder of the mathematics of the infinite, in contrast to whom the inductive ass [1] Newton appears as a plagiariser and corrupter ; Kant, the theory of cosmic evolution *before* Laplace ; Oken, the first in Germany to adopt the theory of evolution ; Hegel, whose [encyclopædic] comprehensive treatment and rational grouping of the natural sciences is a greater achievement than all the materialistic nonsense put together.)

On Büchner's claim to pronounce judgement on socialism and economics from the struggle for existence : Hegel, *Encyclopædia*, I, p. 9, on cobbling.[2]

On politics and socialism : the understanding for which the world has waited, p. 11.[3]

Separation, co-existence, and succession. Hegel, *Encyclopædia*, p. 35 ! [4] As determination of the sensuous, of the idea.

Hegel, *Encyclopædia*, p. 40.[5] Natural phenomena— but in Büchner not *thought out*, merely copied, hence the superfluous.

Page 42.[6] Solon's law, " produced out of his head "—Büchner is able to do the same for modern society.

Page 45.[7] Metaphysics—the science of *things*—not of movements.

Page 53.[8] In experience . . . arrives at

Page 56. The parallelism between the human individual and history = the parallelism between embryology and palæontology.

---

[1] It is impossible to render Engels' word " *Induktionsesel* " into English. A donkey in German idiom may mean a fool, a hard worker, or both. It can thus imply praise and blame at the same time. Probably the implication is that Newton did great work with induction, but was unduly afraid of hypotheses. The phrase might be freely rendered " Newton, who staggered under a burden of inductions."

[2], [3], [4] See Appendix II, p. 337.   [5] See Appendix II, p. 338.
[6], [7], [8] See Appendix p. 339.

Dialectics of Natural Science.—Subject matter—matter in motion.

The different forms and varieties of matter are themselves only to be recognised through motion, only in this are the properties of bodies exhibited ; of a body that does not move there is nothing to be said.[1] Hence the constitution of moving bodies results from the forms of motion.

1. The first, simplest form of motion is the mechanical form, pure change of place :

(*a*) Motion of a single body—does not exist, only relative motion—falling.

(*b*) The motion of separated bodies : trajectory, astronomy—apparent equilibrium—the end always *contact.*

(*c*) The motion of bodies in contact in relation to one another—pressure. Statics. Hydrostatics and gases. The lever and other forms of mechanics proper—which all in their simplest form arise from contact on friction or impact, which are distinct only as stages. But friction and impact, in fact contact, have also other consequences never put forward here by natural scientists : they produce, according to circumstances, sound, heat, light, electricity, magnetism.

2. These different forces (with the exception of sound) —physics of heavenly bodies—

(*a*) pass into one another and mutually replace one another, and

(*b*) on a certain quantitative development of each force applied to the bodies differently in each case, whether they are chemically compound or several chemically simpler bodies, *chemical* changes take place, and we enter the realm of chemistry. Chemistry of heavenly bodies. Crystallographic part of chemistry.

3. Physics had to leave out of account, or was able

[1] This is completely confirmed by modern atomic theory.

to do so, the living organic body ; chemistry finds only in the investigation of organic compounds the real clue to the true nature of the most important bodies, and, on the other hand, it synthesises bodies which only occur in organic nature. Here chemistry leads to organic life, and it has gone far enough to assure us that *it alone* will explain to us the dialectical transition to the organism.

4. The real transition, however, is in *history*—of the solar system, the earth, the *real* pre-condition for organics.

5. Organics.

Divisibility. The mammal is indivisible, the reptile can regrow a foot—ether waves, divisible and measurable to the infinitesimally small—every body divisible, in practice, within certain limits, *e.g.* in chemistry.

Cohesion—already negative in gases—transformation of attraction into *repulsion*, the latter only real in gas and ether (?).

States of aggregation—nodal points where quantitative change is transformed into qualitative.

Secchi [1] and the Pope.

Newtonian attraction and centrifugal force — an example of metaphysical thinking : the problem not solved but *posed* for the first time, and this preached as the solution.[2]—Ditto Clausius' decrease of heat.[3]

[1] Secchi was a Jesuit astronomer of the nineteenth century, one of the first to classify stars by their spectra.

[2] Einstein's general theory of relativity takes us at least a step nearer to the solution. It brings gravitation and centrifugal force together as different examples of an essentially similar phenomenon. It has proved its " this-sidedness " by predicting two new facts which have been observed, namely the deflection of light and the change in its colour by gravitational fields.

[3] Clausius, German nineteenth-century physicist, pointed out that according to existing physical theory other forms of motion would ultimately be converted into heat, and thus would be spread out at a uniform temperature. Thus change of all kinds would come to an end. (See pp. 201-2, 216.)

Laplace's theory presupposes only matter in motion—rotation necessary for all bodies suspended in universal space.

Friction and impact produce an *internal* movement of the bodies concerned, molecular motion, differentiated as warmth, electricity, etc., according to circumstances. *This motion, however, is only temporary : cessante causa cessat effectus.* At a definite stage they all become transformed into a *permanent molecular change*, a *chemical change.*[1]

Causa finalis—matter and its inherent motion. This matter is *no abstraction.* Even in the sun the different substances are dissociated and without distinction in their action. But in the *gaseous sphere of the nebular patch* all substances, although separately present, *become merged in pure matter as such*, only as matter, not acting with its specific properties.[2]

(Moreover even in Hegel the contradiction of *causa efficiens* and *causa finalis* is sublated in reciprocal action.)

. The form of development of natural science, in so far as it thinks, is the *hypothesis.* A new fact is observed which makes impossible the previous mode of explanation of the facts belonging to the same group. From this moment onwards new modes of explanation are required—at first based on only a limited number of facts and observations. Further observational material weeds out these hypotheses, doing away with some and

---

[1] *E.g.* if a match is rubbed lightly it is warmed and then cools down again, if rubbed harder it lights up.

[2] In the sun (save for a few compounds in its outer layers) all matter is dissociated into atoms, and the atoms may lose some electrons. Thus all kinds of matter have the same mechanical properties, those of a hot gas. They can be distinguished by their spectra, that is to say the kinds of light which they give out. In a gaseous nebula even this distinction is lost, except for the infinitesimal fraction of atoms which have at any moment enough energy to radiate.

correcting others, until finally the law is established in a pure form. If one should wait until the material for a law was *in a pure form*, it would mean suspending the process of thought until then and, if only for this reason, the law would never come into being.

The number and succession of hypotheses supplanting one another—given the lack of logical and dialectical education among scientists—easily gives rise to the idea that we cannot know the *essence* of things (Haller and Goethe).[1] This is not peculiar to natural science since all human knowledge develops in a curve which twists many times; and in the historical sciences also, including philosophy, theories displace one another, from which, however, nobody concludes that formal logic, for instance, is nonsense. The last form of this outlook is the " thing in itself." In the first place, this assertion that we cannot know the thing in itself (Hegel, *Encyclopædia*, paragraph 44)[2] passes out of science into fantasy. In the second place, it does not add a word to our scientific knowledge, for if we cannot occupy ourselves with things, they do not exist for us. And, thirdly, it is a mere phrase and is never applied. Taken in the abstract it sounds quite sensible. But suppose one applies it. What would one think of a zoologist who said : a dog *seems* to have four legs, but we do not know whether in reality it has four million legs or none at all ? Or of a mathematician who first of all defines a triangle as having three sides, and then declares that he does not know whether it might not have 25 ? That 2×2 *seems* to be 4 ? But scientists take care not to apply the phrase ' the thing in itself ' in natural science, they permit themselves this only in passing into philosophy. This is the best proof how little seriously they take it and of what little value it is itself. If they did take it seriously, what would be the good of investi-

[1], [2] See Appendix II, p. 339.

gating anything ?   Taken historically the thing would
have a certain meaning : we can only know under the
conditions of our epoch and *as far as these reach.*

The transformation of attraction into repulsion and
vice versa is mystical in Hegel, but in substance he
anticipated by it the scientific discovery that came later.
Even in a gas there is repulsion of the molecules, still
more so in more finely-divided matter, for instance in
the tail of a comet, where it even operates with enormous
force.   Even in this Hegel shows his genius by deriving
attraction as a secondary thing from repulsion as something preceding it : a solar system is only formed by
gradual preponderance of attraction over the originally
prevailing repulsion.   Expansion by heat=repulsion.
The kinetic theory of gases.

The contradictoriness of the thought determinations of
reason : *polarisation.*   Just as electricity, magnetism,
etc., become polarised and move in opposites, so do
thoughts.   Just as in the former it is not possible to
maintain any one-sidedness, and no scientist would
think of doing so, so also in the latter.

For anyone who denies causality every natural law is
a hypothesis, among others also the chemical analysis of
heavenly bodies by means of the prismatic spectrum.
What shallowness of thought to remain at such a
viewpoint !

The thing in itself : Hegel, *Logic,* I, 2, p. 10, also later
a whole section on it :

   " Scepticism did not allow itself to say ' *it is* ' ;
modern idealism (*i.e.* Kant and Fichte) did not allow
itself to regard knowledge as knowledge of the thing
in itself. . . . At the same time, however, scepticism
allowed of multiple determinations of its appearance,

or rather its appearance had the entire manifold wealth of the world as content. Likewise the *appearance* of idealism (*i.e.* what idealism calls appearance) comprehends within it the whole range of these manifold determinations. . . . Hence there may well be no being, no thing or thing in itself at the basis of this content ; *it remains for itself as it is ; it is only translated from being into appearance.*"

Hegel, therefore, is here a much more resolute materialist than the modern natural scientists.

The true nature of the determinations of " essence " is expressed by Hegel himself. *Encyclopædia*, I, paragraph 111, addendum [1] : " In essence everything is *relative* " (*e.g.* positive and negative, which have meaning only in their relation, not each for itself).

The so-called axioms of mathematics are the few thought determinations which mathematics needs for its point of departure. Mathematics is the science of magnitudes ; its point of departure is the concept of magnitude. If defines this lamely and then adds the other elementary determinations of magnitude, not contained in the definition, from outside as axioms, where they appear as unproved, and naturally also as *mathematically* unprovable. The analysis of magnitude would yield all these axiom determinations as necessary determinations of magnitude. Spencer is right in as much as what thus appears to us to be the *self-evidence* of these axioms is *inherited*. They are provable dialectically, in so far as they are not pure tautologies.

Part and whole, for instance, are already categories which become inadequate in organic nature. The ejection of seeds—the embryo and the animal produced

[1] See Appendix II, p. 340.

by birth are not to be conceived as a " part " that is
separated from the "whole," which would give a distorted
treatment. It becomes a part only in a dead body,
Encyclopædia, I, p. 268.[1]

Identity—abstract, $a=a$; and negatively, a is not at
the same time equal and unequal to a—likewise inapplicable in organic nature. The plant, the animal, every
cell is at every moment of its life identical with itself
and yet becoming distinct from itself, by absorption and
excretion of substances, by respiration, by cell formation
and death of cells, by the process of circulation taking
place, in short by a sum of incessant molecular changes
which make up life and the sum total of whose results are
evident to our eyes in the phases of life—embryonic life,
youth, sexual maturity, process of reproduction, old
age, death. This is *apart, moreover, from the evolution of
species*. The further physiology develops, the more
important for it become these incessant, infinitely small
changes,[2] and hence the more important for it also the
consideration of difference *within* identity, and the old
abstract standpoint of formal identity, that an organic
being is to be treated as something simply identical with
itself, as something constant,[3] becomes out of date.
Nevertheless, the mode of thought based thereon, together with its categories, persists. But even in inorganic nature identity as such is non-existent in reality.
Every body is continually exposed to mechanical,
physical, and chemical influences, which are always
changing it and modifying its identity. Only in mathematics—an abstract science which is concerned with
creations of thought, regardless whether reflections of

[1] See Appendix II, p. 340.
[2] These changes are the subject of biochemistry, which is now primarily
concerned, not as in Engels' time with discovering substances in living
beings, but in studying their transformations.
[3] This is well shown by a film of plant growth where a day or a week
is compressed into a minute. We see that the leaf, which we are apt
to regard as a thing given, is a stage in a process.

reality or not—are abstract identity, and its opposition
to difference, in place; and even there they are continually being suspended. Hegel, *Encyclopædia*, I, p. 235.[1]
The fact that identity contains difference within itself is
expressed in *every sentence*, where the predicate is
necessarily different from the subject; the *lily* is a
plant, the *rose* is *red*, where, either in the subject or in
the predicate, there is something that is not covered
by the predicate or the subject. Hegel, *Encyclopædia*,
I, p. 231.[2] That from the outset *identity with itself
requires difference from everything else* as its complement,
is self-evident.

Continual change, *i.e.* abolition of abstract identity
with itself, is also found in so-called inorganic things.
Geology is its history. On the surface, mechanical
changes (denudation, frost), chemical changes (weathering), and, internally, mechanical changes (pressure),
heat (volcanic), chemical (water, acids, binding substances), in great upheavals, earthquakes, etc. The
slate of to-day is fundamentally different from the ooze
from which it is formed, the chalk from the loose micro-
scopic shells that compose it, even more so limestone,
which indeed according to some is of purely organic
origin, and sandstone from the loose sea sand, which
again is derived from the weathering of granite, etc., not
to speak of coal.

Positive and Negative.—Can also be given the reverse
names; in electricity, etc., North and South ditto, if
one reverses this and alters the rest of the terminology
in correspondence, everything remains correct. We can
call West East and East West. The sun rises in the
West, and planets revolve from East to West, etc.,
the names alone are changed. Indeed, in physics we
call the real South pole of the magnet, which is attracted

[1], [2] See Appendix II, p. 341.

by the North pole of the earth's magnetism, the *South* [1] *pole*, and it does not matter.

Life and Death.—Already no physiology is held to be scientific if it does not consider death as an essential factor of life (note, Hegel, *Encyclopædia*, I, p. 152–5),[2] the *negation* of life as being essentially contained in life itself, so that life is always thought of in relation to its necessary result, death, which is always contained in it in germ. The dialectical conception of life is nothing more than this. But for anyone who has once understood this, all talk of the immortality of the soul is done away with. Death is either the dissolution of the organic body, leaving nothing behind but the chemical constituents that formed its substance, or it leaves behind a vital principle, more or less the soul, that then survives *all* living organisms, and not only human beings. Here, therefore, by means of dialectics, simply becoming clear about the nature of life and death suffices to abolish an ancient superstition. Living means dying.

Bad Infinity.—The true was already correctly put by Hegel in *filled* space and time, in the process of nature and history. The whole of nature also is now merged in history, and history is only differentiated from natural history as the evolutionary process of *self-conscious* organisms. This infinite complexity of nature and history has within it the infinity of space and time— bad infinity—only as a sublated moment, essential but not predominant. The extreme limit of our natural science until now has been *our* universe, and we do not need the infinitely numerous universes outside it to obtain knowledge of nature. Indeed, only a single sun among millions, with its solar system, forms the

---

[1] A slip of the pen; it should be *North pole*.
[2] See Appendix II, p. 341.

essential basis of our astronomical researches. For terrestrial mechanics, physics, and chemistry we are more or less restricted to our little earth, and for organic science entirely so. Yet this does not do any essential injury to the practically infinite diversity of phenomena and natural knowledge, any more than history is damaged by the similar, even greater limitation to a comparatively short period and small portion of the earth.

Simple and Compound.—Categories which even in organic nature likewise lose their meaning and become inapplicable. An animal is expressed neither by its mechanical composition from bones, blood, gristle, muscles, tissues, etc., nor by its chemical composition from the elements. Hegel, *Encyclopædia*, I, p. 256.[1] The organism is *neither* simple *nor* compound, however complex it may be.

Primordial matter.—" The conception of matter as originally existent and without form in itself is very old and to be met with even among the Greeks, first of all in the mythical figure of chaos, which is conceived as the formless basis of the existing world." *Encyclopædia*, I, p. 258. We find this chaos again in Laplace, and approximately in the nebula which also has only the *beginning* of form. Differentiation comes afterwards.

The incorrect *theory of porosity* (in which the various false matters, caloric, etc., are situated in the pores of one another and yet do not penetrate one another) is presented by Hegel, *Encyclopædia*, I, p. 259,[2] as a pure *figment of the mind;* see also his *Logic.*

Force.—If any kind of motion is transferred from one body to another, then one can regard the motion, *in so*

[1], [2] See Appendix II, p. 341.

far as it transfers itself, as active, as the cause of the motion, and *in so far as it becomes transferred*, as passive, and then this cause, the active motion, appears as *force* and the passive as its *manifestation*. From the law of the indestructibility of motion, it follows automatically that the force is exactly as great as its manifestation, since indeed it is *the same motion* in both cases. Motion that transfers itself, however, is more or less quantitatively determinable, because it appears in two bodies, of which one can serve as a unit of measurement in order to measure the motion in the other. The measurability of motion gives the category *force* its value, otherwise it has none. Hence the more this is the case, the more are the categories of force and manifestation useful in research. Especially is this so in mechanics, where one resolves the forces still further and regards them as compound, thereby often arriving at new results, although one should not forget that this is merely a mental operation ; by applying the analogy of forces that are really compound, as expressed in the parallelogram of forces, to forces that are really simple, the latter still do not thereby become really compound. Similarly in statics. Then, again, in the transformation of other forms of motion into mechanical motion (heat, electricity, magnetism in the attraction of iron), where the original motion can be measured by the mechanical effect produced. But here, where various forms of motion are considered simultaneously, the limitation of the category or abbreviation, *Force*, already stands revealed. No regular physicist any longer terms electricity, magnetism, or heat mere forces, any more than *substances* or imponderabilia. When we know into how much mechanical motion a definite quantity of heat motion is converted, we still do not know anything of the nature of heat, however much the examination of these transformations may be necessary for the investigation

of the nature of heat. To conceive heat as a form of
motion is the latest advance of physics, and by so doing
the category of force is sublated in it : in certain connections—those of transition—they can appear as forces
and so be measured. Thus heat is measured by the
expansion of a body on warming. If heat did not pass
here from one body to the other—the measuring rod—
i.e. if the heat of the body acting as a measuring rod did
not alter, there could be no talk of measurement, of a
change of magnitude. One says simply : heat expands
a body, whereas to say : heat has the force to expand a
body, would be a mere tautology, and to say : heat is the
force which expands bodies, would not be correct, since
expansion, *e.g.* in gases, is produced also by other means,
and heat is not exhaustively characterised in this way.

Some chemists speak also of chemical force, as the
force that makes and maintains compounds. Here,
however, there is no real transference, but a combination
of the motion of various bodies into a single whole, and
so " force " here reaches its limit. It is, however, still
measurable by the heat production, but so far without
much result. Here it becomes a phrase, as everywhere
where, instead of investigating the unanalysed forms of
motion, one *invents* a so-called force for their explanation
(as, for instance, explaining the floating of wood in
water by a floating force—the refraction of light by a
refractive force, etc.), in which case as many forces are
obtained as there are unexplained phenomena, the
external phenomenon being indeed merely translated
into an internal phrase. (Attraction and repulsion are
easier to excuse ; here a number of phenomena inexplicable to the physicist are embraced under a
common name, which gives an inkling of an inner connection.) Finally in organic nature the category of
force is completely inadequate and yet continually
applied. True, it is possible to characterise the action

of the muscles, in accordance with their mechanical
effect, as muscular force, and also to measure it. One
can even conceive of other measurable functions as
forces, *e.g.* the digestive capacity of various stomachs,
but one quickly arrives *ad absurdum* (*e.g.* nervous force),
and in any case one can speak here of forces only in a
very restricted and figurative sense (the ordinary phrase :
to regain one's forces). This abuse, however, has led
to speaking of a vital force. If by this is meant that
the form of motion in the organic body is different from
the mechanical, physical, or chemical form, and contains
them all sublated in itself, then it is a very lax manner of
expression, and especially so because the force—pre-
supposing transference of motion—appears here as
something pumped into the organism from outside, not as
inherent in it and inseparable from it, and therefore this
vital force has been the last refuge of all supernaturalists.

The defect : (1) Force usually treated as having
independent existence. Hegel, *Naturphil.* [*Philosophy of
Nature*], p. 79.[1]

(2) *Latent, dormant* force—this to be explained from
the relation of motion and rest (inertia, equilibrium),
where also arousing of forces to be dealt with.

The indestructibility of motion in *Descartes'* principle
that *the universe always contains the same quantity of
motion.* Natural scientists express this imperfectly as
the " indestructibility of force." The merely quantitative expression of Descartes is likewise inadequate :
motion as such, as essential activity, the form of existence
of matter, as indestructible as the latter itself, includes the
quantitative element. So here again the philosopher has
been confirmed by the natural scientist after 200 years.

" Its essence (of motion) is to be the immediate unity
of space and time . . . to motion belong space and

―――――――
[1] See Appendix II, p. 342.

time; velocity, the quantum of motion, is space in relation to a definite time that has elapsed." Hegel, *Naturphil*. [*Philosophy of Nature*], p. 65.[1] ". . . Space and time are filled with matter . . . just as there is no motion without matter, so there is no matter without motion," p. 67.[2]

Force (see above).—The transference of motion takes place, of course, only in the presence of *all* the various conditions, which are often very numerous and complex, especially in machinery (the steam engine, the shotgun with lock, trigger, percussion cap, and gunpowder). If one of them is missing, then the transference does not take place until this condition is supplied. One can imagine this as if the force must first be *aroused* by the introduction of this last condition, as if it lay *latent* in a body, the so-called carrier of force (gunpowder, charcoal), whereas in reality not only this body but all the other conditions must be present in order to evoke precisely this special transference.

The notion of force comes to us quite automatically in that we possess in our own body means for transferring motion, which within certain limits can be brought into action by our will; especially the muscles of the arms through which we produce mechanical change of place and motion of other bodies, lifting, carrying, throwing, hitting, etc., resulting in definite useful effects. The motion is here apparently *produced*, not transferred, and this gives rise to the notion of force in general *producing motion*. That muscular force is also merely transference has only recently been proved physiologically.[3]

Motion and equilibrium.—Equilibrium is inseparable

---

[1], [2] See Appendix II, p. 343.

[3] The chemical origin of muscular energy is now understood in much greater detail, and the first few steps in the harder problem of the origin of the energy liberated in the brain have led to important advances in the treatment of insanity.

from motion. In the motion of the heavenly bodies there is *motion in equilibrium* and *equilibrium in motion* (relative). But all specifically relative motion, *i.e.* in this case all separate motion of individual bodies on one of the heavenly bodies in motion, is an effort to establish relative rest, equilibrium. The possibility of a body being at relative rest, the possibility of temporary states of equilibrium, is the essential condition for the differentiation of matter and hence for life. On the sun there is no equilibrium of the various substances, only of the mass as a whole, or at any rate only a very restricted one, determined by considerable differences of density ; on the surface there is eternal motion and unrest, dissociation. On the moon, equilibrium appears to prevail exclusively, without any relative motion— death (moon=negativity). On the earth motion has become differentiated into interchange of motion and equilibrium : the individual motion strives towards equilibrium, the motion as a whole once more destroys the individual equilibrium. The rock comes to rest, but weathering, the action of the ocean surf, of rivers and glacier ice continually destroys the equilibrium. Evaporation and rain, wind, heat, electric and magnetic phenomena offer the same spectacle. Finally, in the living organism we see continual motion of all the smallest particles as well as of the larger organs, resulting in the continual equilibrium of the total organism during the normal period of life, which yet always remains in motion, the living unity of motion and equilibrium.[1] All equilibrium is only *relative* and *temporary*.

Causality.—The first thing that strikes us in considering matter in motion is the interconnection of the

[1] The truth of this statement is constantly being demonstrated afresh. For example, it has been shown that during life even the bones, which appear so solid, are constantly exchanging phosphorus atoms with the blood.

individual motions of separate bodies, their *being
determined* by one another. But not only do we find
that a particular motion is followed by another, we find
also that we can evoke a particular motion by setting
up the conditions in which it takes place in nature,
indeed that we can produce motions which do not occur
at all in nature (industry), at least not in this way, and
that we can give these motions a predetermined direction
and extent. *In this way,* by the *activity of human
beings,* the idea of *causality* becomes established, the
idea that one motion is the *cause* of another. True, the
regular sequence of certain natural phenomena can by
itself give rise to the idea of causality : the heat and
light that come with the sun ; but this affords no proof,
and to that extent Hume's scepticism was correct in
saying that a regular *post hoc* can never establish a
propter hoc. But the activity of human beings *forms
the test* of causality. If we bring the sun's rays to a focus
by means of a lens and make them act like the rays of
an ordinary fire, we thereby prove that the heat comes
from the sun. If we bring together in a rifle the priming,
the explosive charge, and the bullet and then fire it,
we count upon the effect known in advance from previous experience, because we can follow in all its details
the whole process of ignition, combustion, explosion
by the sudden conversion into gas and pressure of the
gas on the bullet. And here the sceptic cannot even
say that because of previous experience it does not follow
that it will be the same next time. For, as a matter of
fact, it does sometimes happen that it is *not* the same,
that the priming or the gunpowder fails to work, that
the barrel bursts, etc. But it is precisely this which
proves causality instead of refuting it, because we can
find out the cause of each such deviation from the rule
by appropriate investigation : chemical decomposition
of the priming, dampness, etc., of the gunpowder, defect

in the barrel, etc., etc., so that here the test of causality is so to say a *double* one.

Natural science, like philosophy, has hitherto entirely neglected the influence of men's activity on their thought [1] ; both know only nature on the one hand and thought on the other. But it is precisely *the alteration of nature by men*, not solely nature as such, which is the most essential and immediate basis of human thought, and it is in the measure that man has learned to change nature that his intelligence has increased. The naturalistic conception of history, as found, for instance, to a greater or lesser extent in Draper and other scientists, as if nature exclusively reacts on man, and natural conditions everywhere exclusively determined his historical development, is therefore one-sided and forgets that man also reacts on nature, changing it and creating new conditions of existence for himself. There is damned little left of " nature " as it was in Germany at the time when the Germanic peoples immigrated into it. The earth's surface, climate, vegetation, fauna, and the human beings themselves have continually changed, and all this owing to human activity, while the changes of nature in Germany which have occurred in the process of time without human interference are incalculably small.

Newtonian Gravitation.—The best that can be said of it is that it does not explain but *pictures* the present state of planetary motion. The motion is given. Ditto the force of attraction of the sun. With these data, how is the motion to be explained ? By the parallelogram of forces, by a tangential force which now becomes a necessary postulate that we *must*

[1] Since Engels' time physicists are beginning to think in terms of operations (human activities) rather than to consider themselves as merely passive observers. But outside physics the tendency has not yet developed appreciably.

accept. That is to say, assuming the *eternal character* of the existing state, we need a *first impulse*, God. But neither is the existing planetary state eternal nor is the motion originally compound, but *simple rotation*, and the parallelogram of forces applied here is wrong, because it did not merely show clearly what was the unknown magnitude, the *x*, that had still to be found, that is to say in so far as Newton claimed not merely to put the question but to solve it.

Force.—The negative side also has to be analysed : the resistance which is opposed to the transference of the motion.[1]

Reciprocal action is the first thing that we encounter when we consider matter in motion as a whole from the standpoint of modern natural science. We see a series of forms of motion, mechanical motion, heat, light, electricity, magnetism, chemical union and decomposition, transitions of states of aggregation, organic life, all of which, if *at present* we *still* make an exception of organic life, pass into one another, mutually determine one another, are in one place cause and in another effect, the sum-total of the motion in all its changing forms remaining the same (Spinoza: *substance is causa sui*—strikingly expresses the reciprocal action). Mechanical motion becomes transformed into heat, electricity, magnetism, light, etc., and *vice versa*. Thus natural science confirms what Hegel has said (where ?) that reciprocal action is the true *causa finalis* of things. We cannot go back further than to knowledge of this reciprocal action, for the very reason that there is nothing behind to know. If we know the forms of motion of matter (for

[1] Here again the progress of physics has been dialectical. The more energy (motion in the broadest sense) a body has, the more its inertia, *i.e.* resistance to being moved. It is possible that all inertia is a manifestation of energy.

which it is true there is still very much lacking, in view of the short time that natural science has existed), then we know matter itself, and therewith our knowledge is complete. (Grove's whole misunderstanding about causality rests on the fact that he does not succeed in arriving at the category of reciprocal action; he has the thing, but not the abstract thought, and hence the confusion—pp. 10–14. [1]) Only from this universal reciprocal action do we arrive at the real causal relation. In order to understand the separate phenomena, we have to tear them out of the general inter-connection and consider them in isolation, and there the changing motions appear, one as cause and the other as effect.

The indestructibility of motion, a pretty passage in Grove, p. 20 *et seq.*[2]

Mechanical Motion.—Among natural scientists motion is always as a matter of course taken as mechanical motion, change of place. This is a survival from the pre-chemical eighteenth century and makes a clear conception of the processes much more difficult. Motion, as applied to matter, is *change in general*. From the same misunderstanding is derived also the craze to reduce everything to mechanical motion—even Grove is " strongly inclined to believe that the other affections of matter . . . are, and will ultimately be resolved into, modes of motion," p. 16 [3]—which obliterates the specific character of the other forms of motion. This is not to say that each of the higher forms of motion is not always necessarily connected with real mechanical (external or molecular) motion, just as the higher forms of motion simultaneously also produce other forms; chemical action is not possible without change of

[1] See Appendix II, p. 343.   [2], [3] See Appendix II, p. 346.

temperature and electric changes, organic life without mechanical, molecular, chemical, thermal, electric, changes, etc. But the presence of these subsidiary forms does not exhaust the essence of the main form in each case. One day we shall certainly " reduce " thought experimentally to molecular and chemical motions in the brain ; but does that exhaust the essence of thought ?

The divisibility of matter. For science the question is in practice a matter of indifference. We known that in chemistry there is a definite limit to divisibility, beyond which bodies can no longer act chemically—the atom ; and that several atoms are always [1] in combination— the molecule. Ditto in physics we are driven to the acceptance of certain—for physical analysis—smallest particles, the arrangement of which determines the form and cohesion of bodies, their vibrations becoming evident as heat, etc. But whether the physical and chemical molecules are identical or different, we do not yet know.[2] Hegel very easily gets over this question of divisibility by saying that matter is both divisible and continuous, and at the same time neither of the two, which is no answer (see sheet 5, 3 below : Clausius), but is now almost proved.

Natural scientific thought.—Agassiz's plan of creation, according to which God proceeded in creation from the general to the particular and individual, first creating the vertebrate as such, then the mammal as such, the animal of prey as such, the cat as such, and only finally

---

[1] This is not strictly correct, though generally believed fifty years ago. A few elements, *e.g.* neon and mercury, exist as single atoms at ordinary temperatures, and all do so when very hot.

[2] We now know that this is true for some substances, but not for all. For example, a metal owes its mechanical properties to the fact that it is built up of very small crystals, each consisting of millions of atoms.

the lion, etc.!   That is to say, first of all abstract ideas
in the shape of concrete things and then concrete things !
See Hæckel, p. 59.[1]

Induction and Deduction. Hæckel, p. 75 *et seq.*,[2]
where Goethe draws the inductive conclusion that man,
who *does not normally have* a premaxillary bone, *must*
have one, hence by *incorrect* induction arrives at something correct.

In Oken (Hæckel, p. 85 *et seq.*) the nonsense that has
arisen from the dualism between natural science and
philosophy is evident.   By the path of thought, Oken
discovers protoplasm and the cell, but it does not occur
to anyone to follow up the matter along the lines of
natural science—it is to be accomplished by *thought!*
And when protoplasm and the cell were discovered,
Oken was in general disrepute !

Causæ finales and efficientes transformed by Hæckel,
pp. 89–90,[3] into *purposively* acting and *mechanically*
acting causes, because for him *causa finalis*=God !
Likewise for him mechanical, simply according to
Kant,=monistic, not=mechanical in the sense of
mechanics.   With such confusion of language, nonsense
is inevitable.   What Hæckel says here of Kant's "*Critique
of Judgment,*" does not agree with Hegel, *G.d. Phil.*
[*History of Philosophy*], p. 603.[4]

God is nowhere treated worse than by natural
scientists, who believe in him. Materialists simply
explain the *facts*, without making use of such phrases,
they do this first when importunate pious believers try
to force God upon them, and then they answer curtly,

---

[1] See Appendix II, p. 346.   [2] See Appendix II, p. 347.
[3], [4] See Appendix II, p. 348.

either like Laplace : *Sire, je n'avais pas, etc.*,[1] or more rudely in the manner of the Dutch merchants who, when German commercial travellers press their shoddy goods on them, are accustomed to turn them away with the words : *Ik kan die Zaken niet gebruiken* [I have no use for the things], and that is the end of the matter. But what God has had to suffer at the hands of his defenders ! In the history of modern natural science, God is treated by his defenders as Frederick William III was treated by his generals and officials in the campaign of Jena. One division of the army after another lowers its weapons, one fortress after another capitulates before the march of science, until at last the whole infinite realm of nature is conquered by science, and there is no place left in it for the Creator. Newton still allowed Him the " first impulse " but forbade Him any further interference in his solar system. Father Secchi bows Him out of the solar system altogether, with all canonical honours it is true, but none the less categorically for all that, and he only allows Him a creative act as regards the primordial nebula. And so in all spheres. In biology, his last great Don Quixote, Agassiz even ascribes positive nonsense to Him ; He is supposed to have created not only the actual animals but also abstract animals, the fish as such ! And finally Tyndall[2] totally forbids Him any entry into nature and relegates him to the world of emotional processes, only admitting Him because, after all, there must be somebody who knows more about all these things (nature) than J. Tyndall ! What a distance from the old God— the Creator of heaven and earth, the maintainer of all things—without whom not a hair can fall from the head !

[1] When Napoleon asked him why God did not appear in his " System of the World," he answered " Sir, I have had no reason to employ that hypothesis."

[2] See Appendix II, p. 349.

Tyndall's emotional need proves nothing. The Chevalier des Grieux also had an emotional need to love and possess Manon Lescaut, who sold herself and him over and over again ; for her sake he became a card-sharper and pimp, and if Tyndall wants to reproach him, he replies with his " emotional need ! "

God=nescio ; but *ignorantia non est argumentum* (Spinoza).

Aggregates in Nature.—Insect states (the ordinary ones do not go beyond purely natural conditions), here even a social aggregate. Ditto productive animals with tools (bees, etc., beavers), but still only subsidiary things and without total effect. Even earlier : colonies of corals and hydrozoa, where the individual is at most an intermediate stage and the fleshly community mostly a stage of the full development. See Nicholson. Similarly, the infusoria, the highest, and in part very much differentiated, form which a single cell can achieve.

Unity of Nature and Mind.—To the Greeks it was self-evident that nature could not be unreasonable, but even to-day the stupidest empiricists prove by their reasoning (however wrong it may be) that they are convinced from the outset that nature cannot be unreasonable or reason contrary to nature.

The classification of sciences, each of which analyses a single form of motion, or a series of forms that belong together and pass into one another, is then the classification, the arrangement, of these forms of motion themselves according to their inherent sequence, and herein lies its importance.

At the end of the last century, after the French materialists who were predominantly mechanical, the need became evident for an *encyclopædic comprehensive treatment* of the entire natural science of the *old* Newton-Linnæus school, and two men of the greatest genius undertook this, Saint Simon (uncompleted) and *Hegel*. To-day, when the new outlook on nature is complete in its basic features, the same need makes itself felt, and attempts are being made in this direction. But where now the general evolutionary connection in nature has to be shown, an external side by side arrangement is as inadequate as Hegel's artificially constructed dialectical transitions. The transitions must make themselves, they must be natural. Just as one form of motion develops out of another, so their reflections, the various sciences, must arise necessarily the one from another.

Protista.[1]—1. Non-cellular, begin with a simple granule of protein which extends and withdraws pseudopodia in one form or another, including the monera. (The Monera of the present day are certainly very different from the original forms, since for the most part they live on organic matter, swallowing diatoms and infusoria, *i.e.* bodies higher than themselves which only arose after them), and, as Hæckel's plate I shows, have a developmental history and pass through the form of non-cellular ciliate swarm-spores. The tendency towards form which characterises all albuminous bodies [2] is already evident here. This tendency is more prominent in the non-cellular foraminifera, which excrete highly artificial shells (anticipating

[1] This whole passage is based on observations which are only partly correct. The development of microscopical technique has shown that the simplest organisms large enough to be visible have a great deal of structure. All the organisms which Engels put in groups 1 and 2 now turn out to have nuclei. On the other hand, some of the ultramicroscopic viruses turn out to be single protein molecules. That is to say they have no structure except the chemical structure that belongs to them as proteins.

[2] Protein molecules may aggregate into crystals, fibres, or what are called tactoids, which resemble organic structures, and of which indeed many organic structures within the cell are examples.

colonies ? corals, etc.) and anticipate the higher
molluscs in form just as the tubular algæ (Siphoneæ)
anticipate the stem, root, and leaf form of higher plants,
although they are merely structureless albumen. Hence
Protamœba is to be separated from *Amœba*.

2. On the one hand there arises the distinction of
skin (ectosarc) and medullary layer (endosarc) in the
sun animalcule *Actinophrys sol,* Nicholson, p. 49. The
skin puts out pseudopodia (in *Protomyxa aurantiaca,*
this stage is already a transitional one, see Hæckel,
plate I). Along this line of evolution protein does not
appear to have got very far.

3. On the other hand, there become differentiated in
the albumen the *nucleus* and *nucleolus*—naked *Amœbæ.*
From now on the development of form proceeds apace.
Similarly, the development of the young cell in the
organism, *cf. Wundt* on this (at the beginning). In
A. sphærococcus, as in *Protomyxa,* the formation of the
cell membrane is only a transitional phase, but even
here there is already the beginning of the circulation
in the contractile vacuole. Sometimes we find either a
shell of sand grains stuck together (*Difflugia,* Nicholson,
p. 47) as in worms and insect larvæ, sometimes a genuinely
excreted shell. Finally,

4. *The cell with a permanent cell membrane.* According
to Hæckel, p. 382,[1] out of this has arisen, depending on
the hardness of the cell membrane, either plant, or in
the case of a soft membrane, animal (it certainly cannot
be conceived so generally ?). With the cell membrane,
definite and at the same time plastic form makes its
appearance. Here again a distinction between simple
cell membrane and excreted shell. But (in contrast to
No. 3) the *putting out of pseudopodia* stops with this cell
membrane and this shell. Repetition of earlier forms
(ciliate swarm-spores) and diversity of form. The

[1] See Appendix II, p. 349.

transition is provided by the *Labyrinthuleæ*, Hæckel, p. 385,[1] which deposit their pseudopodia outside and creep about in this network with alteration of the normal spindle shape kept within definite limits. The *Gregarinæ* anticipate the mode of life of higher *parasites*— some are already no longer single cells but *chains* of cells—Hæckel, p. 451, but only containing 2–3 cells— a poor sort of aggregate. The highest development of unicellular organism is in the *Infusoria*, in so far as these are *really* unicellular. Here a considerable differentiation (see Nicholson). Once again colonies and plant animals (*Epistylis*). Among unicellular plants likewise a high development of form (*Desmidiaceæ*, Hæckel, p. 410).[1]

5. The next advance is the union of several cells into a body, no longer a colony. First of all, the *Katallaktæ* of Hæckel, *Magosphæra planula*, Hæckel, p. 384,[1] where the union of the cells is only a phase in development. But here also there are already no pseudopodia (whether there are any as a transitional phase Hæckel does not state exactly). On the other hand, the *Radiolaria*, also undifferentiated masses of cells, have retained their pseudopodia and have developed to the highest extent the geometric regularity of the shell, which plays a part even among the genuinely non-cellular rhizopods. The protein surrounds itself, so to speak, with its crystalline form.

6. *Magosphæra planula* forms the transition to the true *Planula* and *Gastrula*, etc. Further details in Hæckel, p. 452 *et seq.*[1]

The Individual.—This concept also has been dissolved into something purely relative. Cormus, colony, tapeworm—on the other hand, cell and segment as individuals in a certain sense (anthropogeny and morphology).

[1] See Appendix II, pp. 349-51.

Repetition of morphological forms at all stages of evolution : cell forms (the two essential ones already in Gastrula)—segment formation at a certain stage ; annelids, arthropods, vertebrates. In the tadpoles of amphibians the primitive form of ascidian larvæ is repeated. Various forms of marsupials, which recur among placentals [1] (even counting only existing marsupials).

For the entire evolution of organisms the law of acceleration according to the square of the distance in time from the point of departure is to be accepted. *Cf.* Hæckel, *History of Creation and Anthropogeny*, the organic forms corresponding to the various geological periods. The higher, the more rapidly it proceeds.

The whole of organic nature is one continuous proof of the identity or inseparability of form and content. Morphological and physiological phenomena, form and function, mutually determine one another. The differentiation of form (the cell) determines differentiation of substance into muscle, skin, bone, epithelium, etc., and the differentiation of material in turn determines difference of form.

The Kinetic theory of gases : " In a perfect gas . . . the molecules are already so far distant from one another than their mutual interaction can be neglected " (Clausius, p. 6).[2] *What fills up the spaces between them?* Ditto ether. Hence here *the postulate of a matter that is not made up of molecular or atomic cells.*

The law of identity in the old metaphysical sense is the fundamental law of the old outlook : a=a. Each thing is equal to itself. Everything was permanent, the solar system, stars, organisms. This law has been refuted by natural science bit by bit in each separate

[1] *E.g.* there are marsupials closely resembling the dog and the mole.
[2] See Appendix II, p. 851.

case, but theoretically it still prevails and is still put
forward by the supporters of the old in opposition to
the new : a thing cannot simultaneously be itself and
something else.   And yet the fact that true, concrete
identity includes difference, change, has recently been
shown in detail by natural science (see above).   Abstract
identity, like all metaphysical categories, suffices for
everyday use, where small-scale conditions or brief
periods of time are in question ; the limits within which
it is usable differ in almost every case and are determined
by the nature of the object.   For a planetary system,
where for ordinary astronomical calculation the ellipse
can be taken as the basic form without committing errors
in practice, they are much wider than for an insect that
completes its metamorphosis in a few weeks. (Give
other examples, *e.g.* alteration of species, which is
reckoned in periods of many thousands of years.)   For
natural science in its comprehensive role, however,
even in each single branch, abstract identity is totally
insufficient, and although on the whole it has now been
abolished in practice, theoretically it still dominates
people's minds, and most natural scientists imagine that
identity and difference are irreconcilable opposites,
instead of one-sided poles the truth of which lies only in
their reciprocal action, in the inclusion of difference
within identity.

Natural scientists believe that they free themselves
from philosophy by ignoring it or abusing it.   They
cannot, however, make any headway without thought,
and for thought they need thought determinations.
But they take these categories unreflectingly from the
common consciousness of so-called educated persons,
which is dominated by the relics of long obsolete philo-
sophies, or from the little bit of philosophy compulsorily
listened to at the university (which is not only frag-

mentary, but also a medley of views of people belonging
to the most varied and usually the worst schools), or
from uncritical and unsystematic reading of philosophical writings of all kinds. Hence they are no less
in bondage to philosophy, but unfortunately in most
cases to the worst philosophy, and those who abuse
philosophy most are slaves to precisely the worst
vulgarised relics of the worst philosophers.

Historical.—Modern natural science—the only one
which comes in question *qua* science as against the
brilliant intuitions of the Greeks and the sporadic
unconnected investigations of the Arabs—begins with
that mighty epoch when feudalism was smashed by the
burghers. In the background of the struggle between
the burghers of the towns and the feudal nobility this
epoch showed the peasant in revolt, and behind the
peasant the revolutionary beginnings of the modern
proletariat, already red flag in hand and with com-
munism on its lips. It was the epoch which brought
into being the great monarchies in Europe, broke the
spiritual dictatorship of the Pope, evoked the revival
of Greek antiquity and with it the highest artistic
development of the new age, broke through the boundaries of the old world, and for the first time really discovered the world.

It was the greatest revolution that the world had so far
experienced. Natural science also moved and had its
being in this revolution, was revolutionary through and
through, advanced hand in hand with the awakening
modern philosophy of the great Italians, and provided
its martyrs for the stake and the prisons. It is characteristic that protestants and catholics vied with one
another in persecuting it. The former burned Servetus,
the latter Giordano Bruno. It was a time that called
for giants and produced giants, giants in learning,

intellect, and character, a time that the French correctly
called the Renaissance and protestant Europe with one-
sided prejudice called the time of the Reformation.

At that time natural science also had its declaration
of independence, though it is true it did not come right
at the beginning, any more than that Luther was the
first protestant. What Luther's burning of the papal
bull was in the religious field, in the field of natural
science was the great work of Copernicus, in which he,
although modestly, after thirty years' hesitation and
so to say on his death bed, threw down a challenge to
ecclesiastical superstition. From then on natural science
was in essence emancipated from religion, although the
complete settlement of accounts in all details has gone
on to the present day and in many minds is still far from
being complete. But from then on the development of
science went forward with giant strides, increasing, so
to speak, proportionately to the square of the distance
in time from its point of departure, as if it wanted to
show the world that for the motion of the highest
product of organic matter, the human mind, the law of
inverse squares holds good, as it does for the motion of
inorganic matter.

The first period of modern natural science ends—in the
inorganic sphere—with Newton. It is the period in
which the available subject matter was mastered ; it
performed a great work in the fields of mathematics,
mechanics and astronomy, statics, and dynamics,
especially owing to Kepler and Galileo, from whom
Newton drew his conclusions. In the organic sphere,
however, there had been no progress beyond the first
beginnings. The investigation of the forms of life
historically succeeding one another and replacing one
another, as well as the changing conditions of life
corresponding to them—palæontology and geology—
did not yet exist. Nature was not at all regarded as

something that developed historically, that had a history in time ; only extension in space was taken into account ; the various forms were grouped not one after the other, but only one beside the other ; natural history was valid for all periods, like the elliptical orbits of the planets. For any closer analysis of organic structure both the immediate bases were lacking, *viz.* chemistry and knowledge of the essential organic structure, the cell. Natural science, at the outset revolutionary, was confronted by an out-and-out conservative nature, in which everything remained to-day as it was at the beginning of the world, and in which right to the end of the world everything would remain as it had been in the beginning.

It is characteristic that this conservative outlook on nature both in the inorganic and in the organic sphere . . .[1]

Astronomy		Geology
Mechanics	Physics	Palæontology
Mathematics	Chemistry	Mineralogy

Plant physiology	
Animal physiology	Therapeutics
Anatomy	Diagnostics

The first breach : Kant and Laplace. The second : Geology and Palæontology (Lyell, slow development). The third : organic chemistry, which prepares organic bodies and shows the validity of chemical laws for living bodies. The fourth : 1842, mechanical heat, Grove. The fifth : Darwin, Lamarck, the cell, etc. (Struggle, Cuvier and Agassiz). The sixth : the *comparative element* in anatomy, climatology (isotherms), scientific expeditions since the middle of the eighteenth century, animal and plant geography, physical geo-

---

[1] The sentence is incomplete in the manuscript—*Ed.*

graphy in general (Humboldt). The assembling of the material in its interconnection. Morphology (embryology, Baer).

The old teleology has gone to the devil, but the certainty now stands firm that matter in its eternal cycle moves according to laws which at a definite stage —now here, now there—necessarily give rise to the thinking mind in organic beings.

The normal existence of animals is given by the conditions in which they live and to which they adapt themselves—those of man, as soon as he differentiates himself from the animal in the narrower sense, have as yet never been present, and are only to be elaborated by the ensuing historical development. Man is the sole animal capable of working his way out of the merely animal state—his normal state is one appropriate to his consciousness, *one to be created by himself*.

The contradictory character of theoretical development ; from the *horror vacui* the transition was made at once to absolutely empty space, only afterwards the *ether*.

Generatio Æquivoca.[1]—All investigations hitherto as follows : in fluids containing organic matter in decomposition and accessible to the air, lower organisms arise, protista, fungi, infusoria. Whence do they come ? Have they arisen by *generatio œquivoca*, or from germs brought in from the atmosphere ? Consequently the investigation is limited to a quite narrow field, to the question of plasmogony.

The assumption that new living organisms can arise by the decomposition of others belongs essentially to the

---

[1] This phrase is generally translated as " spontaneous generation." The whole section which follows is extremely up-to-date. Except for a few details, Engels' argument holds good to-day, and a great many facts have been discovered which confirm it.

epoch of immutable species. At that time men found themselves compelled to assume the origin of all organisms, even the most complicated, by original generation from non-living materials, and if they did not want to resort to the aid of an act of creation, they easily arrived at the view that this process is more readily explicable given a formative material already derived from the organic world ; no one any longer believes in the production of a mammal directly from inorganic matter by chemical means.

This assumption, however, directly conflicts with the present state of science. By the analysis of the process of decomposition in dead organic bodies chemistry proves that at each successive step this process necessarily produces products that more and more approximate to the dead inorganic world, products that are less and less capable of being used by the organic world, and that this process can be given another direction, such utilisation being able to occur only when these products of decomposition are absorbed early enough in an appropriate, already existing, organism. It is precisely the most essential vehicle of cell-formation, protein, that decomposes first of all, and so far it has never been built up again.

Still more. The organisms whose original generation from organic fluids is the question at issue in these investigations, while being of a comparatively low order, are nevertheless definitely differentiated, bacteria, yeasts, etc., with a life cycle composed of various phases and in part, as in the case of the infusoria, equipped with fairly well developed organs. They are all at least unicellular. But ever since we have been acquainted with the structureless Monera,[1] it has become foolish

[1] The " Monera " are not structureless. Actually the gap between protozoa and bacteria on the one hand, and small filter-passing viruses on the other, is larger than the gap which Engels mentions. His argument has in fact been strengthened.

to desire to explain the origin of even a single cell directly from dead matter instead of from structureless living protein, to believe it is possible by means of a little stinking water to force nature to accomplish in twenty-four hours what it has cost her thousands of years to bring about.

Pasteur's attempts in this direction are useless; for those who believe in this possibility he will never be able to prove their impossibility by these experiments alone, but they are important because they furnish much enlightenment on these organisms, their life, their germs, etc.

Force.—Hegel, *Gesch. d. Phil.* [*History of Philosophy*], I, p. 208, says :

> " It is better expressed (as Thales does) that the magnet has a *soul*, than that it has an attracting *force* ; force is a sort of property that, *separable from matter*, is imagined as a predicate—soul, on the other hand, *being its movement, identical with the nature of matter*."

Hæckel, *Anthropology*, p. 707.[1] " According to the materialist outlook on the world, *matter or stuff* was present **earlier than motion** or *vis viva*, matter created force." This is just as false as that force created matter, since force and matter are inseparable. Where does he get his materialism from ?

Mayer, *Mechanische Theorie der Wärme* [*Mechanical Theory of Heat*], p. 328. *Kant has already stated* that the ebb and flow of tides exert a retarding pressure on the rotating earth. (Adams' calculation that the duration of the sidereal day [2] is now increasing by 1/100 second in 1,000 years).

[1] See Appendix II, p. 352.

[2] *I.e.* the time between two successive passages of the same " fixed " star across the meridian at a given point. This is much more nearly constant than the ordinary or solar day, and can also of course be measured far more accurately.

An example of the necessity of dialectical thought and of non-rigid categories and relations in nature; the law of falling, which already in the case of a period of fall of some minutes becomes incorrect, since then the radius of the earth can no longer without error be put $=\infty$, and the attraction of the earth increases instead of remaining constant as Galileo's law of falling assumes. Nevertheless, this law is still continually taught, but the reservation omitted!

Moriz Wagner, *Naturwissenschaftliche Streitfragen* [*Controversial Questions of Natural Science*], I (*Augsburger Allgemeine Zeitung, Beilage,* 6, 7, 8, October, 1874).

Liebig's statement to Wagner in his last year, 1868 :

" We may only assume that life is just as old and just as eternal as matter itself, and the whole controversial point about the origin of life seems to me to be disposed of by this simple assumption. In point of fact, why should not organic life be thought of as present from the very beginning just as much as carbon and *its compounds* (!), or as the whole of uncreatable and indestructible matter in general, and the forces that are eternally bound up with the motion of matter in space ? "

Liebig said further (Wagner believes November, 1868) that he also regards the hypothesis, that organic life has been " imported " on to our planet from universal space, as " acceptable."

Helmholtz (Preface to Thomson's *Handbuch der theoretischen Physik* [*Handbook of Theoretical Physics*], German edition, part II) :

" It appears to me to be a fully correct procedure, *if all our efforts fail to cause the production of organisms from non-living substance*, to raise the question whether life has ever arisen, whether it is not just as old as matter, and whether its germs have not been transported from one heavenly body to another and have developed wherever they have found favourable soil."

Wagner :

" The fact that matter is indestructible and imperishable, that it . . . can by no force be reduced to nothing, *suffices for the chemist to regard it as ' uncreatable '* . . . But according to the now prevailing view (?) life is regarded merely as a ' property ' inherent in certain simple elements, of which the lowest organisms consist, and which, as a matter of course, must be as old, *i.e.* as originally existing, as these basic stuffs *and their compounds* (! !) themselves."

In this sense one could also speak of vital force, as Liebig does (*Chemische Briefe* [*Letters on Chemistry*], 4th edition) :

" Namely as ' a formative principle in and together with the physical forces,' hence not acting outside of matter.   This vital force as a ' property of matter,' however, manifests itself . . . only under appropriate conditions which have existed since eternity at innumerable points in infinite space, but which in the course of the different periods of time must often enough have been spatially varied."

Hence no life is possible on the ancient fluid earth or the present-day sun, but the glowing bodies have enormously expanded atmospheres, consisting, according to recent views, of the same materials that fill all space in extremely rarified form and are attracted by bodies. The rotating nebular mass from which the solar system developed, reaching beyond the orbit of Neptune, contained " also all water (!) dissolved in vaporous form in an atmosphere richly impregnated with carbonic acid (!) up to immeasurable heights, and with that also the basic materials for the existence (?) of the lowest organic germs"; in it there prevailed " the most various degrees of temperature in the very different regions, and hence the assumption is *fully justified* that at all times the conditions necessary for organic life were somewhere to be found. According to this the atmospheres of the heavenly bodies like those of the rotating cosmic nebular masses, would have to be regarded as the permanent repositories of the living form, as the eternal breeding grounds of organic germs." In the Andes, below the equator, the smallest living protista with their invisible germs are still present in masses in the atmosphere up to 16,000 feet. Perty says that they are " almost omnipresent." They are only absent where the glowing heat kills them. Hence their existence (*Vibrionidæ*, etc.) is conceivable " also in the vapour belt of *all* heavenly bodies, wherever the appropriate conditions are to be found."

" According to Cohn, bacteria are . . . so extremely minute that 633 million can find room in a cubic millimetre, and 636,000 million weigh only a gramme. The micrococci are even smaller," and perhaps they are not the smallest. But being very varied in shape, " the *Vibrionidæ* . . . sometimes globular, sometimes ovoid, sometimes rod-shaped or spiriform," (already possess, therefore, a considerable measure of form). " Hitherto no valid objection has been raised against the well-founded hypothesis that all the multifarious, more highly organised living beings of both natural kingdoms *could* have developed and *must* have developed in the course of very long periods of time from such, *or similar*, extremely simple (! !), neutral, primordial beings, hovering between plants and animals . . . on the basis of individual variability and the capacity for hereditary transmission of newly acquired characters to the offspring on alteration of the physical conditions of the heavenly bodies and on spatial separation of the individual varieties produced."

The proof, worth noting, how much of a dilettante

Liebig was in biology, although the latter is a science bordering on chemistry. He read Darwin for the first time in 1861, and only much later the important biological and palæontological-geological works subsequent to Darwin. Lamarck he had " never read." " Similarly the important palæontological special researches which appeared even before 1859, of L. V. Buch, d'Orbigny, Münster, Klipstein, Hauer, and Quenstadt on the fossil Cephalopods, that throw such remarkable light on the genetic connection of the various creations, remained completely unknown to him. All the above-mentioned scientists were . . . driven by the force of facts, almost against their will, to the Lamarckian hypothesis of descent," and this indeed *before* Darwin's book. The theory of descent, therefore, had already quietly struck roots in the views of those scientists who had concerned themselves more closely with the comparative study of fossil organisms. . . . As early as 1832, in " On the Ammonites and their Division into Families," L. V. Buch very definitely introduced in the science of petrefacts (!) " The Lamarckian Idea of the Typical Relationship of Organic Forms as a Sign of their Common Descent," the title of a paper read before the Berlin Academy in 1848, and in 1848 he based himself on his investigation of the ammonites for the declaration : " that the disappearance of old forms and the appearance of new ones is not a consequence of the total destruction of the organic creations, but that *the formation of new species out of older forms has most probably only resulted from altered conditions of life.*"

Comments.—The above hypothesis of " eternal life " and of importation presupposes :

1. The eternal existence of protein.

2. The eternal existence of the original forms from which everything organic can develop. Both are inadmissible.

Ad. 1.—Liebig's assertion, that carbon compounds are just as eternal as carbon itself, is inexact, if not false.

(*a*) Is carbon simple ? [1] If not, it is as such not eternal.

(*b*) The compounds of carbon are eternal in the sense that under similar conditions of mixture, temperature, pressure, electric potential, etc., they always reproduce themselves. But that, for instance, only the simplest carbon compounds, CO_2 or CH_4, should be eternal in the sense that they exist at all times and more or less in all places, and not rather that they are continually produced anew and pass out of existence again—in fact, out of the elements and into the elements—has hitherto not been asserted. If living protein is eternal in the same sense as other carbon compounds, then it must not only continually be dissolved into its elements, as notoriously happens, but also continually be produced anew from the elements and without the collaboration of previously existing protein—and that is the exact opposite of the result at which Liebig arrives.

(*c*) Protein is the most unstable carbon compound known to us. It decomposes as soon as it loses the capacity of carrying out the functions peculiar to it, which we call life, and it is inherent in its nature that this incapacity should sooner or later make its appearance. And it is just this compound which is supposed to be eternal and able to endure all the changes of temperature, pressure, lack of nourishment, and air, etc., in space, although even its upper temperature limit is so low—less than $100°$ C. The conditions for the existence of protein are infinitely more complicated than those of any other known carbon compound, because not only

---

[1] It is noteworthy that Engels questioned the eternity of the chemical elements. It is now, of course, a commonplace that they can be transformed, and it is at least conceivable that all carbon has been formed from hydrogen and neutrons.

physical and chemical functions, but in addition
nutritive and respiratory functions, enter, requiring a
medium which is narrowly delimited, physically and
chemically—and is it this medium that one must
suppose has maintained itself from eternity under all
possible changes ? Liebig " prefers, *ceteris paribus*, the
simpler of two hypotheses," but a thing may appear
very simple and yet be very complicated. The assumption of innumerable continuous series of living protein
bodies, tracing their descent from one another through
all eternity, and which under all circumstances always
leave sufficient over for the stock to remain well assorted,
is the most complicated assumption possible. Moreover, the atmospheres of the heavenly bodies, and
especially nebular atmospheres, were originally glowing
hot and therefore no place for protein bodies—hence
in the last resort space must serve as the great reservoir,
a reservoir in which there is neither air nor nourishment,
and with a temperature at which certainly no protein
can function or maintain itself !

Ad. 2.—The vibrios, micrococci, etc., of which we
are speaking, are beings already considerably differentiated—protein granules [1] that have excreted an outer
membrane, *but no nucleus*. The series of protein bodies
capable of development, however, *forms a nucleus first
of all* and becomes a cell –the cell membrane is then
a further advance (*Amœba sphærococcus*). Hence the
organisms with which we have been dealing here belong
to a series which, by all previous analogy, proceeds
barrenly into a blind alley, and they cannot be numbered
among the ancestors of the higher organisms.

What Helmholtz says of the sterility of attempts to
produce life artificially is pure childishness. Life is

---

[1] These organisms include other substances (*e.g.* fats and waxes)
besides proteins. The argument is however quite correct if applied to
the smaller viruses.

the mode of existence of protein bodies, the essential element of which consists in *continual, metabolic interchange with the natural environment outside them*, and which ceases with the cessation of this metabolism, bringing about the decomposition of the protein.[1] If success is ever attained in preparing protein bodies chemically, they will exhibit the phenomena of life and carry out metabolism, however weak and short-lived they may be.[2] But it is certain that such bodies could *at most* have the form of the very crudest monera, and probably much lower forms, but by no means the form of organisms that have become differentiated by an evolution lasting thousands of years, and in which the cell membrane has become separated from the contents and a definite inherited form assumed. So long, however, as we know no more of the chemical composition of protein than we do at present, and therefore for probably another hundred years to come cannot think of its artificial preparation, it is ridiculous to complain that all our efforts, etc., have failed !

Against the above assertion that metabolism is the characteristic activity of protein bodies may be put the objection of the growth of Traube's " artificial cells." But here there is merely unaltered absorption of a liquid by endosmosis, while metabolism consists in the absorption of substances, the chemical composition of which is altered, which are assimilated by the organism, and the residua of which are excreted together with

[1] Such metabolism can also occur in the case of inorganic bodies and in the long run it occurs everywhere, since chemical reactions take place, even if extremely slowly, everywhere. The difference, however, is that inorganic bodies are destroyed by this metabolism, while in organic bodies it is the necessary condition for their existence. [*Note by F. Engels.*]

[2] We now doubt whether all proteins would do so. A number have been isolated, though none have yet been made from their elements. Some, however, carry out some of the processes of life. Hæmoglobin absorbs and takes up oxygen, pepsin digests other proteins, virus nucleoprotein even reproduces itself in a favourable environment.

the decomposition products of the organism itself resulting from the life process.[1] The significance of Traube's " cells " lies in the fact that they show endosmosis and growth as two things occurring also in inorganic nature and which can be exhibited without any carbon.

The newly arisen protein granule must have had the capacity of nourishing itself from oxygen, carbon dioxide, ammonia, and some of the salts dissolved in the surrounding water. Organic nutritive substances were not present,[2] for the granules surely could not devour one another. This proves how high above them are the present-day monera, even without nuclei, living on diatoms, etc., and therefore presupposing a number of differentiated organisms.

Reaction.—Mechanical, physical (alias heat, etc.) reaction is exhausted with each occurrence of reaction. Chemical reaction alters the composition of the reacting body and is only renewed if a further quantity of the latter is added. Only the *organic* body reacts *independently*, of course within its sphere of power (sleep), and assuming the supply of nourishment—but this supply of nourishment is effective only after it has been assimilated, not immediately, as at lower stages, so that here the organic body has an *independent* power of reaction, the new reaction must be *brought about* by it.

---

[1] *N.B.*—Just as we have to speak of invertebrate vertebrates, so also here the unorganised, formless, undifferentiated granule of protein is termed an organism—*dialectically* this is permissible because just as the vertebral column is implicit in the notochord so in the protein granule on its first origin the whole infinite series of higher organisms lies included " *in itself* " as if in an embryo. [*Note by F. Engels.*]

[2] It now seems likely that simple organic substances were present in the primordial ocean, synthesised by the ultra-violet rays of the sunlight, which in the absence of oxygen and ozone were less absorbed in the upper atmosphere. These would not have decayed in the absence of bacteria, and could therefore have served as food for the first living (or semi-living) things.

Identity and Difference.—The dialectical relation is already seen in the differential calculus, where dx is infinitely small, but yet is effective and performs everything.

Mathematics.—Nothing appears more solidly based than the difference of the four species, the elements of all mathematics. Yet right at the outset multiplication is seen to be an abbreviated addition, and division an abbreviated subtraction, of a definite number of equal numerical magnitudes ; and in one case—when the divisor is a fraction—division is even carried out by multiplying by the inverted fraction. In algebraic calculation the thing is carried much further. Every subtraction $(a-b)$ can be represented as an addition $(-b+a)$, every division $\dfrac{a}{b}$ as a multiplication $a\dfrac{1}{b}$. In calculations with powers of magnitudes one goes much further still. All rigid differences between the kinds of calculation disappear, everything allows of being presented in the opposite form. A power can be put as a root $(x^2=\sqrt{x^4})$, a root as a power $(\sqrt{x}=x^{\frac{1}{2}})$. Unity divided by a power or root can be put as a power of the denominator $\left(\dfrac{1}{\sqrt{x}}=x^{-\frac{1}{2}}\,;\,\dfrac{1}{x^3}=x^{-3}\right)$. Multiplication or division of the powers of a magnitude becomes converted into addition or subtraction of their exponents. Any number can be conceived and expressed as the power of any other number (logarithms, $y=a^x$). And this transformation of one form into the opposite one is no idle trifling, it is one of the most powerful levers of mathematical science, without which to-day hardly any of the more difficult calculations are carried out. If negative and fractional powers alone were abolished from mathematics, how far could one get ?

$(-\cdot-=+,\div=+,\sqrt{-1}$, etc., to be expounded earlier.$)$

The turning point in mathematics was Descartes'
variable magnitude. With that came *motion* and hence
dialectics in mathematics, and *at once also of necessity
the differential and integral calculus*, which moreover
immediately begins, and which on the whole was perfected
by Newton and Leibniz, not discovered by them.

Asymptotes.—Geometry begins with the discovery that
straight and curved are absolute opposites, that straight
is absolutely inexpressible in curved, and curved in
straight, that the two are incommensurable. Yet even
the calculation of the circle [1] is only possible by expressing its periphery in straight lines. For curves
with asymptotes, however, straight becomes totally
merged in curved, and curved in straight ; just as much
as the notion of parallelism : the lines are not parallel,
they continually approach one another and yet never
meet ; the arm of the curve becomes more and more
straight, without ever becoming entirely so, just as in
analytical geometry the straight line is regarded as a
curve of the first order with an infinitely small curvature.
The x of the logarithmic curve [2] may become ever so
large, y can never $=0$.

Zero Powers.—Of importance in the logarithmic series :
$$\begin{array}{cccc} 0 & 1 & 2 & 3 \\ 10^0 & 10^1 & 10^2 & 10^3 \end{array} \text{ log}$$
All variables pass somewhere
through unity ; hence also constants raised to a variable
power $a^x=1$, if $x=o$. $a^o=1$ means nothing more than
the conception of unity in its connection with the other
members of the series of powers of a, only there has it
any meaning and can lead to results [3] $\left(\sum x^o = \dfrac{\omega}{x} \right)$,

---

[1] *I.e.* the expression of its circumference in terms of its radius.
[2] *I.e.* the rectangular hyperbola $xy=c$.
[3] The expression in the bracket is meaningless as it stands. However Engels' writing is by no means easy to decipher and we cannot
be sure what Engels actually wrote.

otherwise not at all. From this it follows that unity
also, however much it may appear identical with itself,
includes within it an infinite manifoldness, since it can
be the zero power of any other possible number, and
that this manifoldness is not merely imaginary is proved
on each occasion that unity is conceived as a particular
unity, as one of the variable results of a process (as a
momentary magnitude or form of a variable) in connection with this process.

Straight and curved in the differential calculus [1] are in
last resort put as equal : in the differential triangle, the
hypotenuse of which forms the differential of the arc (in
the tangent method), this hypotenuse can be regarded
" comme une petite ligne tout droite qui est tout a la
fois l'élément de l'arc et celui de la tangente "—if now
the curve is regarded as composed of an infinite number
of straight lines, or also, however, " lorsqu'on la considère
comme rigoureuse ; puisque le détour à chaque point
M étant infiniment petit, la raison dernière de l'élément
de la courbe à celui de la tangente *est evidemment une
raison d'égalité.*" Here, therefore, although the ratio
continually *approaches* equality, but *asymptotically* in
accordance with the nature of the curve, yet, since the
contact is limited to a single *point* which has no length,
it is finally assumed that equality of straight and curved
has been reached. Bossut, *Calcul. diff. et intégr.*
[*Differential and Integral Calculus*], Paris, An. VI, I,
p. 149. In polar curves the differential imaginary
abscissæ [2] are even taken as parallel to the real abscissæ
and operations based on this, although both meet at
the pole ; indeed, from it is deduced the equality of two
triangles, one of which has an angle precisely at the

[1] This was, of course, written before " rigorous " proofs based on
the theory of limits were introduced into most books on the calculus.
Engels is quite correct concerning the calculus as taught in his day.
[2] In modern terminology radii vectores.

point of intersection of the two lines, the parallelism of which is the whole basis of the equality ! Fig. 17.

When the mathematics of straight and curved lines has thus pretty well reached exhaustion a new almost infinite field is opened up by the mathematics that *conceives curved as straight* (the differential triangle) and *straight as curved* (curve of the first order with infinitely small curvature).   O metaphysics !

Ether.[1]—If the ether offers resistance at all, it must also offer resistance to *light*, and so at a certain distance be impenetrable to light.   That however ether *propagates* light, being its *medium*, necessarily involves that it should also offer resistance to light, otherwise light could not set it in vibration.   This the solution of the controversial questions raised by Mädler [2] and mentioned by Lavrov.[2]

Vertebrates.—Their essential character : the *grouping of the whole body about the nervous system.* Thereby the development to self-consciousness, etc., becomes possible.   In all other animals the nervous system is a secondary affair, here it is the basis of the whole organisation ;   the nervous system, when developed to a certain extent—by posterior elongation of the head ganglion of the worms—takes possession of the whole body and organises it according to its needs.

Radiation of Heat into Interstellar Space.—All the hypotheses cited by Lavrov of the renewal of extinct heavenly bodies (p. 109) [3] *involve loss of motion.*   The heat once radiated, *i.e.* the infinitely greater part of the

---

[1] Few physicists now believe in the ether as they did fifty years ago. The notion had to be abandoned when it was shown that motion of bodies relative to it could not be detected.   Engels' note, therefore, has validity only as a comment on the physical ideas of his time.

[2] See Appendix II, p. 352.

[3] See Appendix II, p. 353.

original motion, is and remains lost. Helmholtz says, up to now, 453/454. Hence one finally arrives after all at the exhaustion and cessation of motion. The question is only finally solved when it has been shown how the heat radiated into space becomes *utilisable* again. The theory of the transformation of motion puts this question categorically, and it cannot be evaded by extending the period of operation or by evasion. That, however, with the posing of the question the conditions for its solution are simultaneously given—*c'est autre chose*. The transformation of motion and its indestructibility were first discovered hardly thirty years ago, and it is only quite recently that they have been further analysed and followed up in regard to their consequences. The question as to what becomes of the apparently lost heat has, as it were, only been *nettement posée* since 1867 (Clausius). No wonder that it has not yet been solved; it may still be a long time before we arrive at a solution with our small means. But it will be solved, just as surely as it is certain that there are no miracles in nature and that the original heat of the nebular ball is not communicated to it miraculously from outside the universe. The general assertion that *the amount of motion is infinite*, and hence inexhaustible, is of equally little assistance in overcoming the difficulties of each individual case; it too does not suffice for the revival of extinct universes, except in the cases provided for in the above hypothesis, which are always bound up with loss of force and are therefore only temporary cases. The cycle [1] has not been traced and will not be until the re-utilisation of the radiated heat shall have been discovered.

Newton's Parallelogram of Forces in the solar system is true at any rate *for the moment when the annular bodies*

[1] See note to p. 24.

separate, because then the rotational motion comes into contradiction with itself, appearing on the one hand as attraction, and on the other hand as tangential force. As soon as the separation is complete, however, the motion is again a unity. That this separation must occur is a proof for the dialectical process.

Bathybius.[1]—The stones in its flesh are proof that the original form of protein, still lacking any differentiation of form, already bears within it in germ the capacity for skeletal formation.

Understanding and Reason.—This Hegelian distinction, according to which only dialectical thinking is reasonable, has a definite meaning. All activity of the understanding we have in common with animals : *induction, deduction,* and hence also *abstraction* (Dido's [2] generic concepts : quadrupeds and bipeds), *analysis* of unknown objects (even the cracking of a nut is a beginning of analysis), *synthesis* (in animal tricks), and, as the union of both, *experiment* (in the case of new obstacles and unfamiliar situations). In their nature all these modes of procedure—hence all means of scientific investigation that ordinary logic recognises—are absolutely the same in men and the higher animals. They differ only in degree (of development of the method in each case). The basic features of the method are the same and lead to the same results in man and animals, so long as both operate or make shift merely with these elementary methods.

On the other hand, dialectical thought—precisely because it pre-supposes investigation of the nature of concepts—is only possible for man, and for him only at a comparatively high stage of development (Buddhists and Greeks), and it attains its full development much

[1] See Appendix II, p. 353.   [2] Engels had a dog called Dído.

later still through modern philosophy—and yet we have the colossal results already among the Greeks (!) which go far in anticipating investigation.

Chemistry, in which *analysis* is the predominant form of investigation, is nothing without its complementary pole : *synthesis*.

To the Pan-Inductionists.—With all the induction in the world we would never have got to the point of becoming clear about the *process* of induction. Only the *analysis* of this process could accomplish this. Induction and deduction belong together as necessarily as synthesis and analysis. Instead of one-sidedly raising one to the heavens at the cost of the other, one should seek to apply each of them in its place, and that can only be done by bearing in mind that they belong together, that each completes the other. According to the inductionists, induction would be an infallible method. It is so little so that its apparently surest results are everyday overthrown by new discoveries. Light corpuscles, caloric, were results of induction. Where are they now ? Induction taught us that all vertebrates have a central nervous system differentiated into brain and spinal cord, and that the spinal cord is enclosed in cartilaginous or bony vertebræ —whence indeed the name is derived. Then *Amphioxus* was revealed as a vertebrate with an undifferentiated central nervous strand and *without* vertebræ. Induction established that fishes are those vertebrates which throughout life breathe exclusively by means of gills. Then animals come to light whose fish character is almost universally recognised, but which, besides gills, have also well-developed lungs, and it turns out that every fish carries a potential lung in the swim bladder. Only by audacious application of the theory of evolution did Hæckel rescue the inductionists, who were feeling quite comfortable in these contradictions. If induction were really so infallible, whence come the rapid successive revolutions in classification of the organic world ? They are the most characteristic product of induction, and yet they annihilate one another.

The Kinetic theory has to show [1] how molecules that strive upwards can at the same time exert a downwards pressure and—assuming the atmosphere as more or less permanent in relation to interstellar space—how in spite of gravity they can move to a distance from the centre of the earth, but nevertheless, at a certain distance, although the force of gravity has decreased according to the *square* of the distance, are yet compelled by this force to come to a stop or to return.

Clausius—if correct—proves that the universe has been created, *ergo* that matter is creatable, *ergo* that it is destructible, *ergo* that also force, or motion, is creatable and destructible, *ergo* that the whole theory of the " conservation of force " is nonsense, *ergo* that all its consequences are also nonsense.

The notion of an actual *chemically uniform matter—* ancient as it is—fully corresponds to the childish view, widely held even up to Lavoisier, that the chemical affinity of two bodies depends on each one containing a common third body (Kopp, *Entwicklung*, p. 105).[2]

Hard and fast lines are incompatible with the theory of evolution. Even the border line between vertebrates and invertebrates is now no longer rigid, just as little

---

[1] This has now been accomplished.
[2] See Appendix II, p. 354.

is that between fishes and amphibians, while that
between birds and reptiles dwindles more and more
every day. Between *Compsognathus* and *Archæopteryx* [1] only a few intermediate links are wanting, and
birds' beaks with teeth crop up in both hemispheres.
" Either this—or that ! " becomes more and more
inadequate. Among lower animals the concept of the
individual cannot be established at all sharply. Not
only as to whether a particular animal is an individual
or a colony, but also where in development one
individual ceases and the other begins (nurses).[2]

For a stage in the outlook on nature where all differences become merged in intermediate stages, and all
opposites are bridged by intermediate links, the old
metaphysical method of thought no longer suffices.
Dialectics, which likewise knows no hard and fast
lines, no unconditional, universally valid " either—or ! "
which bridges the fixed metaphysical differences, and
besides " either—or ! " recognises also in the right place
" both this—and that ! " and reconciles the opposites,
is the sole method of thought appropriate in the highest
degree to this stage. For everyday use, for the small
change of science, the metaphysical categories retain
their validity.

Dialectics, so-called *objective* dialectics, prevails
throughout nature, and so-called subjective dialectics,
dialectical thought, is only the reflex of the movement
in opposites which asserts itself everywhere in nature,
and which by the continual conflict of the opposites

[1] *Compsognathus*, a bird-like reptilian fossil ; *Archæopteryx*, a
fossil bird with teeth, a long bony tail, and claws on its wings.
[2] *E.g.* cells or organs whose function is to nourish another cell or
organ. Some parts of the placenta (after-birth) are of maternal and
some of fœtal origin. Is the endosperm of a maize grain which serves
as food for the embryo to be regarded as a separate individual ?
Probably, since it can inherit characters differing from those of the
embryo.

and their final merging into one another, or into higher forms, determines the life of nature. Attraction and Repulsion. Polarity begins with magnetism, it is exhibited in one and the same body ; in the case of electricity it distributes itself over two or more bodies which become oppositely charged. All chemical processes reduce themselves to processes of chemical attraction and repulsion. Finally, in organic life the formation of the cell nucleus is likewise to be regarded as a polarisation of the living protein material, and from the simple cell onwards, the theory of evolution demonstrates how each advance up to the most complicated plant on the one side, and up to man on the other, is effected by the continual conflict between heredity and adaptation. In this connection it becomes evident how little applicable to such forms of evolution are categories like " positive " and " negative." One can conceive of heredity as the positive, conservative side, adaptation as the negative side that continually destroys what has been inherited, but one can just as well take adaptation as the creative, active, positive activity, and heredity as the resisting, passive, negative activity. But just as in history progress makes its appearance as the negation of what exists, so here also—on purely practical grounds—adaptation is better conceived as negative activity. In history, motion in opposites is most markedly exhibited in all critical epochs of the foremost peoples. At such moments a people has only the choice between the two horns of a dilemma : " either —or ! " and indeed the question is always put in a way quite different from that in which the philistines, who dabble in politics in every age, would have liked it put. Even the liberal German philistine of 1848 found himself in 1849 suddenly, unexpectedly, and against his will confronted by the question : a return to the old reaction in an intensified form, or continuance of the revolution up to the republic, perhaps even the one and indivisible republic with a socialist background. He did not spend long in reflection and helped to create the Manteuffel reaction as the flower of German liberalism. Similarly, in 1851, the French bourgeois when faced with the dilemma which he certainly did not expect : a caricature of empire, pretorian rule, and the exploitation of France by a gang of scoundrels, or a social-democratic republic—and he bowed down before the gang of scoundrels so as to be able, under their protection, to go on exploiting the workers.

The Struggle for Life.—Until Darwin, what was stressed by his present adherents was precisely the harmonious co-operative working of organic nature, how the plant kingdom supplies animals with nourishment and oxygen, and animals supply plants with manure, ammonia, and carbonic acid. Hardly was Darwin recognised before these same people saw everywhere nothing but *struggle*. Both views are justified within narrow limits, but both are equally one-sided and prejudiced. The interaction of dead natural bodies includes both harmony and collisions, that of living bodies conscious and unconscious co-operation equally with conscious and unconscious struggle. Hence, even in regard to nature, it is not permissible one-sidedly to inscribe only " struggle " on one's banners. But it is absolutely childish to desire to sum up the whole manifold wealth of historical evolution and complexity in the meagre and one-sided phrase " struggle for life." That says less than nothing.

The whole Darwinian theory of the struggle for life is simply the transference from society to organic nature of Hobbes' theory of *bellum omnium contra omnes*, and of the bourgeois economic theory of competition, as well as the Malthusian theory of population. When once this feat has been accomplished (the unconditional justification for which, especially as regards the Malthusian theory, is still very questionable), it is very easy to transfer these theories back again from natural history to the history of society, and altogether too naive to maintain that thereby these assertions have been proved as eternal natural laws of society.

Let us accept for a moment the phrase " struggle for life " for argument's sake. The most that the animal can achieve is to *collect*; man *produces*, he prepares the means of life in the widest sense of the words, which, without him, nature would not have produced. This makes impossible any immediate transference of the laws of life in animal societies to human ones. Production soon brings it about that the so-called struggle for existence no longer turns on pure means of existence, but on means for enjoyment and development. Here— where the means of development are socially produced— the categories taken from the animal kingdom are already totally inapplicable.[1] Finally, under the capitalist mode of production, production reaches such a height that society can no longer consume the means of life, enjoyment, and development that have been produced, because for the great mass of producers access to these means is artificially and forcibly barred; and therefore every ten years a crisis restores the equilibrium by destroying not only the means of life, enjoyment, and development that have been produced, but also a great part of the productive forces themselves. Hence the so-called struggle for existence assumes the form : to *protect* the products and productive forces produced by bourgeois capitalist society against the destructive, ravaging effect of this capitalist social order, by taking control of social production and distribution out of

---

[1] In particular the struggle ceases to be a Darwinian struggle for life. Even allowing for their lower infant mortality, the bourgeoisie breed more slowly than the workers, and if they win the struggle for wealth, lose the struggle for life.

the hands of the ruling capitalist class, which has become incapable of this function, and transferring it to the producing masses—and that is the socialist revolution.

Even by itself the conception of history as a series of class struggles is much richer in content and deeper than merely reducing it to weakly distinguished phases of the struggle for existence.

Light and darkness are certainly the most conspicuous and definite opposites in nature; they have always served as a rhetorical phrase, from the time of the fourth Gospel to the *lumières* of religion and philosophy in the eighteenth century. Fick, p. 9 : " the law long ago rigidly demonstrates in physics . . . that the form of motion called radiant heat is identical in all essential respects with the form of motion that we call *light*." Clerk Maxwell, p. 14 [1] : " These rays (of radiant heat) have all the physical properties of rays of light and are capable of reflection, etc. . . . some of the heat-rays are identical with the rays of light, while other kinds of heat-rays make no impression upon our eyes."

Hence there exist *dark* light-rays,[2] and the famous opposition between light and darkness disappears from natural science in its absolute form. Incidentally, the deepest darkness and the brightest, most glaring, light have the same effect of *dazzling* our eyes, and so *for us* also they are identical.

The fact is, the sun's rays have different effects according to the length of the vibration, those with the greatest wave-length communicate heat, those with medium wave-length, light, and those with the shortest wave-length,[3] chemical action (Secchi, p. 632 *et seq.*),

[1] See Appendix II, p. 354.
[2] Now generally called infra-red rays, as they have slower frequencies than red light.
[3] *I.e.* ultra-violet radiation. X-rays, of course, have a still shorter wave-length.

the maxima of the three actions being closely approximated, the *inner* minima of the outer group of rays, as regards their action coinciding within the light ray group.[1] What is light and what is non-light depends on the structure of the eye. Night animals may be able to see even a part, not of the heat rays, but of the chemical rays, since their eyes are adapted for shorter wave-lengths than ours. The difficulty disappears if one assumes, instead of three kinds, only a single kind of ray (and scientifically we know only *one* and everything else is a premature conclusion), which has different, but within narrow limits compatible, effects according to the wave-length.[2]

Work.—The mechanical theory of heat has transferred this category from economics into physics (for *physiologically* it is still a long way from having been scientifically determined), but in so doing it becomes defined in quite a different way, as seen even from the fact that only a very slight, subordinate part of economic work (lifting of loads, etc.) can be expressed in kilogram-metres. Nevertheless, there is an inclination to re-transfer the thermodynamical definition of work to the sciences from which the category was derived, with a different determination. For instance, without further ado, to identify it wholesale with physiological work, as in Fick and Wislicenus' Faulhorn experiment,[3] in which the lifting of a human body, say 60 kgs., to a height of say 2,000 metres, *i.e.* 120,000 kilogram-metres, is supposed to express the *physiological* work done. In the physiological work done, however, it makes an

[1] *I.e.* visible rays have slight heating and chemical effects.

[2] This is, of course, correct. There is a continuous series of rays from radio to γ-rays, in which a quantitative change in the wave-length shows itself in great qualitative differences. But this has only been discovered since Engels' death.

[3] These two physiologists observed their metabolism (by collecting urine, etc.) when climbing this mountain.

enormous difference *how* this lifting is effected : whether
by positive lifting of the load by mounting vertical
ladders, or whether along a road or stair with 45° slope
(militarily impracticable ground), or along a road with
a slope of 1/18, hence a length of about 36 kms. (but this
is questionable, if the same time is allowed in all cases).
At any rate, however, in all practicable cases a forward
motion also is combined with the lifting, and indeed
where the road is quite level this is fairly considerable
and as physiological work it cannot be put equal to zero.
In some places there even appears to be not a little
desire to re-import the thermodynamical category of
work back into economics [1] (as with the Darwinists and
the struggle for existence), the result of which would be
nothing but nonsense. Let someone try converting
any skilled labour into kilogram-metres and then
determining wages by means of it ! Physiologically
considered, the human body contains organs which
in their totality, *from one aspect*, can be regarded as a
thermodynamical machine, where heat is supplied and
converted into motion.[2] But even if one pre-supposes
constant conditions as regards the other bodily organs,
it is questionable whether physiological work done,
even lifting, can be at once fully expressed in kilogram-
metres, since within the body internal work is performed
at the same time which does not appear in the result.[3]
For the body is not a steam engine, which only undergoes
friction and depreciation. Physiological work is only
possible with continued chemical changes in the body
itself, depending also on the process of respiration and
the work of the heart. Along with every muscular

[1] Compare Professor Soddy's writings in our own time.

[2] This was commonly believed in Engels' time, but is now known to
be untrue. Chemical energy is not converted into heat before being
transformed into the energy of muscular motion.

[3] Thus an isolated muscle may have an efficiency of nearly 50 per
cent., *i.e.* convert nearly half the available chemical energy into work,
but the efficiency of the body as a whole rarely rises to 25 per cent.

contraction or relaxation, chemical changes occur in the nerves and muscles, and these changes cannot be treated as parallel to those of coal in a steam engine. One can very well compare two instances of physiological work that have taken place under otherwise identical conditions, but one cannot measure the physical work of a man after the manner of that of a steam engine, etc. ; their external results, yes, but not the processes themselves without considerable reservations. (All this has to be greatly revised.)[1]

Induction and Analysis.—A striking example of how little induction can claim to be the sole or even the predominant form of scientific discovery occurs in thermodynamics : the steam engine provided the most striking proof that one can impart heat and obtain mechanical motion. 100,000 steam engines do not prove this more than one, but only more and more forced the physicists into the necessity of providing an explanation. Sadi Carnot [2] was the first seriously to set about the task. But not by induction. He studied the steam engine, analysed it, and found that in it the process which mattered does not appear *in pure form* but is concealed by all sorts of subsidiary processes. He did away with these subsidiary circumstances that have no bearing on the essential process, and constructed an ideal steam engine (or gas engine), which it is true is as little capable of being realised as, for instance, a geometrical line or surface, but in its way performs the same service as these mathematical abstractions : it presents the process in a pure, independent, and unadulterated form. And he came right up against the mechanical equivalent of heat (see the significance of his function C),

[1] Even after sixty years it needs very little revision on the scientific side, except that accurate figures could be given proving the correctness of all but one of Engels' statements.

[2] French physicist of the early nineteenth century.

which he only failed to discover and see because he believed in *caloric*. Here also proof of the damage done by false theories.

The successive development of the separate branches of natural science should be studied. First of all, *astronomy*, which, if only on account of the seasons, was absolutely indispensable for pastoral and agricultural peoples. Astronomy can only develop with the aid of *mathematics*. Hence this also had to be tackled. Further, at a certain stage of agriculture and in certain regions (raising of water for irrigation in Egypt), and especially with the origin of towns, big building operations, and the development of handicrafts—*mechanics*. This was soon needed also for *navigation* and *war*. Moreover, it requires the aid of mathematics and so promotes the latter's development. Thus, from the very beginning the origin and development of the sciences has been determined by production.

Throughout antiquity, scientific investigation proper remained restricted to these three branches, and indeed in the form of exact, systematic research it occurs for the first time in the post-classical period (the Alexandrines, Archimedes, etc.). In physics and chemistry, which were as yet hardly separated in men's minds (theory of the elements, absence of the idea of a chemical element), in botany, zoology, human and animal anatomy, it had only been possible until then to collect facts and arrange them as systematically as possible. Physiology was sheer guesswork, as soon as one went beyond the most tangible things—*e.g.* digestion and excretion—and it could not be otherwise when even the circulation of the blood was not known. At the end of the period, chemistry makes its appearance in its primitive form of alchemy.

If, after the dark night of the Middle Ages was over,

the sciences suddenly arose anew with undreamt-of force, developing at a miraculous rate, once again we owe this miracle to—production. In the first place, following the crusades, industry developed enormously and brought to light a quantity of new mechanical (weaving, clock-making, milling), chemical (dyeing, metallurgy, alcohol), and physical (lenses) facts, and this not only gave enormous material for observation, but also itself provided quite other means for experimenting than previously existed, and allowed the construction of *new* instruments ; it can be said that really systematic experimental science had now become possible for the first time. Secondly, the whole of West and Middle Europe, including Poland, now developed in a connected fashion, even though Italy was still at the head in virtue of its old-inherited civilisation. Thirdly, geographical discoveries—made purely on behalf of gain and, therefore, in the last resort, of production—opened up an infinite and hitherto inaccessible amount of material of a meteorological, zoological, botanical, and physiological (human) bearing. Fourthly, there was the *printing press*.

Now—apart from mathematics, astronomy, and mechanics which were already in existence—physics becomes definitely separate from chemistry (Torricelli, Galileo—the former in connection with industrial waterworks studied first of all the movement of fluids, see Clerk Maxwell). Boyle put chemistry on a stable basis as a science, Harvey did the same for physiology (human and animal) by the discovery of the blood circulation. Zoology and botany remain at first collecting sciences, until palæontology appeared on the scene—Cuvier—and shortly afterwards came the discovery of the cell and the development of organic chemistry. Therewith comparative morphology and physiology became possible and from then on both are true sciences. Geology was founded at the end of the last century, and recently anthropology, badly so-called, enabling the transition from morphology and physiology of man and human races to history. This to be studied further in detail and to be developed.

Clausius' second law,[1] etc., however it may be formulated, shows energy is lost, qualitatively [2] if not quantitatively. *Entropy cannot be destroyed by natural means but it can certainly be created.* The world clock has to be wound up, then it goes on running until it arrives at a state of equilibrium from which only a miracle can set it going again. The energy expended in winding has disappeared, at least qualitatively, and can only be restored by an *impulse from outside.* Hence, an impulse from outside was necessary at the beginning also, hence, the quantity of motion, or energy, existing in the universe was not always the same, hence, energy has been artificially created, *i.e.* it must be creatable, and therefore destructible. *Ad absurdum!*

Difference between the situation at the end of the ancient world, ca. 300—and at the end of the Middle Ages—1453 :

1. Instead of a thin strip of civilisation along the coast of the Mediterranean, stretching its arms sporadically in the interior and up to the Atlantic coast of Spain, France, and England, which could thus easily be broken through and rolled back by the Germans and Slavs from the North, and by the Arabs from the South-East, there was now a closed area of civilisation—the whole of West Europe with Scandinavia, Poland, and Hungary as advance posts.

[1] See Appendix II, p. 355.
[2] *I.e.* other forms of energy are degraded to heat, and high-temperature heat to low-temperature heat.

2. Instead of the contrast between the Greeks, or Romans, and the barbarians, there are now six civilised peoples with civilised languages, not counting the Scandinavian, etc., all of whom had developed to such an extent that they could participate in the mighty rise of literature in the fourteenth century, ensuring a far more diversified culture than that of the Greek and Latin languages, which were already in decay and dying out at the end of ancient times.

3. An infinitely higher development of industrial production and trade, created by the burghers of the Middle Ages ; on the one hand production more perfected, more varied and on a larger scale, and on the other commerce much stronger, navigation being infinitely more enterprising since the time of the Saxons, Friesians, and Normans, and on the other hand also the amount of inventions and importation of oriental inventions, which not only for the first time made possible the importation and diffusion of Greek literature, the maritime discoveries, and the bourgeois religious revolution, but also gave them a quite different and quicker range of action.   In addition they produced a mass of scientific facts, although as yet unsystematised, such as antiquity never had (the magnetic needle, printing, type, flax paper, used by the Arabs and Spanish Jews since the twelfth century, cotton paper gradually making its appearance since the tenth century, and already more widespread in the thirteenth and fourteenth centuries, papyrus quite obsolete in Egypt since the Arabs)—gunpowder, *lenses, mechanical clocks*, great progress both of *chronology* and of *mechanics*.

(See below concerning inventions.)

In addition material provided by travels (Marco Polo, *ca.* 1272, etc.).

General education, even though still bad, much more widespread owing to the universities.

With the rise of Constantinople and the fall of Rome, antiquity comes to an end. The end of the Middle Ages is indissolubly linked with the fall of Constantinople. The new age begins with the return to the Greeks. Negation of the negation !

Historical Material.—Inventions.

B.C.

Fire-hose, water-clock, *ca.* 200 B.C.   Street paving (Rome).

Parchment, *ca.* 160.

A.D.

Water-mill *on the Moselle, ca.* 340, in Germany in the time of Charles the Great.

First signs of glass windows, street lighting in Antioch, *ca.* 370.

Silk-worms from China, *ca.* 550 in Greece.

Pens in the sixth century.

Cotton paper from China to the Arabs in the seventh century, in the ninth in Italy.

Water organs in France in the eighth century.

Silver mines in the Harz worked since the tenth century.

Windmills about 1000.

Notes, Guido of Arezzo's musical scale, *ca.* 1000.

Sericulture introduced in Italy, *ca.* 1100.

Clocks with wheels—ditto.

Needle magnet from the Arabs to the Europeans, *ca.* 1180.

Street paving in Paris 1184.

Lenses in Florence. Glass mirrors.   ⎫ Second half of
Striking clocks, cotton paper in France. ⎬ thirteenth
Herring-salting. Sluices.   ⎭ century.

Rag-paper—beginning of fourteenth century.

Bills of exchange—middle of ditto.

First paper mill in Germany (Nuremberg) 1390.

Street lighting in London. Beginning of fifteenth century.

Post in Venice—ditto.

Wood-cuts and printing—ditto.

Copper-engraving—middle ditto.

Horse post in France 1464.

Silver mines in the Saxon Erzgebirge, 1471.

Harpsichord invented 1472.

Pocket watches. Air-guns. Flintlock—end of fifteenth century.

Spinning wheel 1530.

Diving bell 1538.

Dialectics of Nature.—References.

Nature No. 294 *et seq.* Allman on Infusoria. Unicellular character, important. Croll on **Ice Period and** geological time.

Nature No. 326, Tyndall on *generatio.* Specific decay and fermentation experiments.

Mädler, *Fixed Stars.*

Halley, at the beginning of the eighteenth century, from the difference between the data of Hipparchus and Flamsteed on three stars, first gave the idea of proper motion, p. 410. Flamsteed's British Catalogue, the first approximately accurate and comprehensive one, p. 420, then *ca.* 1750, Bradley, Maskelyne, and Lalande.

Crazy theory of the range of light rays in the case of enormous bodies and Mädler's calculation based on this —as crazy as anything in Hegel's *Philosophy of Nature*, pp. 424–5.

The strongest (apparent) proper motion of a star— $701''$ in a century $=11'$ $41''=$ one-third of the sun's diameter; smallest average of 921 telescopic stars $8\cdot65''$, some $4''$. Milky Way a series of rings, all with a common centre of gravity, p. 434.

The Pleiades Group, and in it Alcyone.—η Tauri, the centre of motion for our island universe " as far as the most remote regions of the Milky Way," p. 448. Period of revolution within the Pleiades group on the average *ca.* two million years, p. 449. About the Pleiades are annular groups alternately poor in stars and rich in stars. Secchi contests the possibility of fixing a centre as the present time.

According to Bessel, *Sirius* and *Procyon* describe an orbit about a dark body, as well as the general motion, p. 450.

Eclipse of Algol every 3 days, duration 8 hours, *confirmed by spectral analysis*, Secchi, p. 786.

In the region of the *Milky Way*, but deep within it, a dense ring of stars of magnitudes 7–11 ; a long way outside this ring are the concentric Milky Way rings, of which we see two. In the Milky Way, according to Herschel, 18 million stars visible through his telescope, those lying within the ring being *ca.* 2 million or more, hence over 20 million in all. In addition there is always a non-resolvable glow in the Milky Way, even behind the resolved stars, hence perhaps still further rings concealed owing to perspective ? Pp. 451–2.

Alcyone distant from the sun 573 light years.[1] *Diameter of the Milky Way ring* of separate visible stars, at least 8,000 light years, pp. 462, 463.

The *mass* of the heavenly bodies moving within the sun-Alcyone radius of 573 light years is calculated at 118 million sun masses, p. 462, not at all in agreement with the at most 2 million stars moving therein. Dark bodies ? At any rate something wrong. A proof how imperfect our observational bases still are.

For the outermost ring of the Milky Way, Mädler assumes a distance of thousands, perhaps of hundreds of thousands, of light years, p. 464.

[1] Mädler's figure, incorrect.

A *beautiful argument* against the so-called absorption of light :

"At any rate, there does exist a distance (from which no further light can reach us), but the reason is quite a different one. The velocity of light is *finite* ; from the beginning of creation to our day a *finite* time has elapsed, and therefore we can only become aware of the heavenly bodies up to the distance which light has traversed in this finite time ! " (p. 466).

That light, decreasing in intensity according to the square of the distance, must reach a point where it is no longer visible to our eyes, however much the latter may be strengthened and equipped, is quite obvious, and suffices [1] for refuting the view of Olbers that only light absorption is capable of explaining the darkness of space that nevertheless is filled in all directions with shining stars to an infinite distance. This is not intended to mean that there does not exist a distance at which the ether *allows no further light to penetrate.*

Nebulæ.—Of all forms, strictly circular, elliptical, or irregular and jagged. All degrees of resolvability, merging into total non-resolvability, where only a thickening towards the centre can be distinguished. In some of the resolvable nebulæ, up to ten thousand stars are perceptible, the middle mostly denser, very rarely a central star of greater brilliance. Rosse's giant telescope has, however, resolved many of them. Herschel I [2] counts 197 star aggregations and 2,300 nebulæ, to which must be added those catalogued by

---

[1] Engels' argument is incorrect here. If space were evenly filled with stars which had been shining for ever as brightly as those in our neighbourhood, and there were no absorption, we should be roasted by starlight !

[2] These refer to the two Herschels, father and son, both first-rate astronomers.

Herschel II in the southern heavens. The irregular ones *must be distant island universes*,[1] since masses of vapour can only exist in equilibrium in globular or ellipsiodal form. Most of them, moreover, are only just visible even through the strongest telescopes. At any rate the circular ones *can* be vapour masses; there are 78 of them among the above 3,500. Herschel assumes 2 million, Mädler—on the assumption of a true diameter equal to 8,000 light years—30 million light years distant from us. Since the distance of any astronomical system of bodies from the next one amounts to at least a hundredfold the diameter of the system, the distance of our universe from the next one would be at least 50 times 8,000 light years $=400,000$ light years, in which case with the several thousands of nebulæ we get far beyond Herschel I's 2 million, p. 492.

Secchi : The resolvable [2] nebulæ give a continuous and an ordinary stellar spectrum. The nebulæ proper, however, " in part give a continuous spectrum like the nebulæ in Andromeda, but usually they give a spectrum consisting of one or only very few bright lines, like the nebulæ in Orion, in Sagittarius, in Lyra, and the majority of those that are known by the name of *planetary* (circular) nebulæ," p. 787. (The nebula in Andromeda according to Mädler, p. 495, is unresolvable —the nebula in Orion is irregularly flocculent and, as it were, puts out arms, p. 495. Those of Lyra and the Cross are only slightly elliptical, p. 498.) Huggins found in the spectrum of Herschel's nebulæ No. 4374, three bright lines, " from this it follows immediately that this nebula does not consist of an aggregate of separate stars, but is a *true nebula*, a glowing substance

---

[1] *I.e.* systems of stars like our own Milky Way. Modern figures, of course, differ appreciably from those given, but are of the same general order.

[2] *I.e.* bodies which appear as nebulæ with a small telescope, but as clusters of stars with a large one.

in the *gaseous* state." The lines belong to nitrogen (I) and hydrogen (I), the third is unknown. Similarly for the nebula in Orion. Even nebulæ that contain gleaming points (Hydra, Sagittarius), have these bright lines, so that star masses in course of aggregation are still not solid or liquid, p. 789. The nebula in Lyra has only a nitrogen line, p. 789. The densest place of the nebula in Orion is 1°, its whole extension 4°.

Secchi : *Sirius :*

"Eleven years later (according to Bessel's calculation, Mädler, p. 450) . . . not only was the satellite of Sirius discovered in the form of a self-luminous star of the sixth magnitude, but it was also shown that its orbit coincides with that calculated by Bessel. Since then the orbit also for Procyon and its companion has been determined by Auwers, although the satellite itself has not yet been seen " (p. 793).

Secchi : *Fixed Stars :* "Since the fixed stars, with the exception of two or three, have no perceptible parallax, they are at least " some 30 light years distant from us, p. 799. According to Secchi, the stars of the 16th magnitude (still distinguishable in Herschel's big telescope) are 7,560 light years distant, those distinguishable in Rosse's telescope are at least 20,900 light years distant, p. 802.

Secchi, p. 810, himself asks : If the sun and the whole system becomes frozen ; "are there forces in nature which can put the dead system back into the original state of glowing nebula and re-awaken it to new life ? We do not know."

The transformation of quantity into quality = " mechanical " world outlook, quantitative change alters quality.   The gentlemen never suspected that !

Identity and difference—necessity and chance—cause and effect—the two main opposites which, treated separately, become transformed into one another.
And then " first principles " must help.

Just as Fourier [1] is a mathematical poem and yet still used, so Hegel a dialectical poem.

Hegel's conception of force and its expression, cause and effect as identical, is proved by the change of form of matter, where the equivalence is shown mathematically. This had already been recognised in measurement. Force is measured by its expression, cause by effect.

The evolution of a concept, for instance, or of a conceptual relation (positive and negative, cause and effect, substance and accident) in the history of thought, is related to its development in the mind of the individual dialectician, just as the evolution of an organism in palæontology is related to its development in embryology (or rather in history and in the single embryo). That this is so was first discovered by Hegel for concepts. In historical development, chance plays its part, which in dialectical thinking, as in the development of the embryo, *is comprised in necessity.*
Abstract and concrete.   The general law of the change of form of motion is much more concrete than any single " concrete " example of it.

[1] Fourier's mathematical theory of heat, in which he founded modern harmonic analysis.   Many of his theorems, as stated, were false. But they were not only beautiful, but of great practical value.   They have now been stated in their correct form, or at least more nearly so.

The significance of names. In organic chemistry the significance of a body, hence also its name, is no longer determined merely by its composition, but rather by its position in the *series* to which it belongs. If we find, therefore, that a body belongs to such a series, its old name becomes an obstacle to understanding it and must be replaced by a *series name* [1] (paraffins, etc.).

Hæckel's Nonsense.—Induction against deduction. As if it were not the case that deduction = inference, and therefore induction also a deduction. This comes from polarisation.

By induction it was discovered 100 years ago that crabs and spiders were insects and all lower animals were worms. By induction it has now been found that this is nonsense and there exist x classes. Wherein then lies the advantage of the so-called inductive conclusion, which can be just as false as the deductive conclusion, the basis of which is classification ?

Induction can never prove that there will never be a mammal without lacteal glands. Formerly nipples were the mark of a mammal. But the platypus has none,

The whole swindle about induction was invented by the Englishmen ; Whewell, *Inductive Sciences*, comprising the purely mathematical side, and so the contradiction to deduction. Logic, old or new, knows nothing of this. All forms of conclusion that start from single things are experimental and based on experience, indeed the inductive conclusion even starts from A—E—B (general).

[1] The organic chemists attempted to carry this out at a congress held at Geneva in 1892. Thus valeric acid (originally so-called because it was made from valerian) may be called pentanoic acid to show that it can be derived from the five-carbon paraffin pentane by oxidising its terminal carbon atom.

It is also characteristic of the thinking capacity of our natural scientists that Hæckel fanatically champions induction at the very moment when the *results* of induction—the systems of classification—are everywhere put in question (*Limulus* [1] a spider, *Ascidia* [2] a vertebrate or chordate, the *Dipnoi*,[3] however, being fishes, in opposition to all original definitions of amphibia) and daily new facts are being discovered which overthrow the *entire* previous classification by induction. What a beautiful confirmation of Hegel's thesis that the inductive conclusion is essentially a problematic one ! [4] Indeed, even the whole classification of organisms has been taken away from induction owing to the theory of evolution, and referred back to " deduction," to heredity —one species being literally *deduced* from another by heredity—and it is impossible to prove the theory of evolution by induction alone, since it is quite anti-inductive.   The concepts with which induction operates : species, genus, class, have been rendered fluid by the theory of evolution and so have become *relative* : but one cannot use relative concepts for induction.

Induction   and   Deduction.   Hæckel, *History   of Creation*, pp. 76–7.   The   conclusion   polarised   in induction and deduction !

Polarisation.—For J. Grimm it was still a firmly established law that a German dialect must be either High German or Low German.   In this he totally lost sight of the Frankish dialect. Because the written Frankish of the later Carolingian period was High German (since the High German shifting of consonants

---

[1] The king-crab, shown by Marx's friend Ray Lankester to be an arachnid, *i.e.* related to the spiders and scorpions, though not, of course, exactly a spider.

[2] A sea-squirt. Though the adult is sessile, the larva resembles a tadpole.

[3] Lungfishes.

[4] See Appendix II, p. 355.

had taken possession of the Frankish South-East), he imagined that Frankish passed in one place into old High German, in another place into French. It then remained absolutely impossible to explain the source of the Netherland dialect in the ancient Salic regions. Frankish was only rediscovered after Grimm's death : Salic in its rejuvenation as the Netherland dialect, Ripuaric in the Middle and Lower Rhine dialects, which in part have been shifted to various stages of High German, and in part have remained Low German, so that Frankish is a dialect that is *both* High German *and* Low German.

Polarity.—A magnet, on being cut through, polarises the neutral middle portion, but in such a way that the old poles remain. On the other hand a worm, on being cut into two, retains the receptive mouth at the positive pole and forms a new negative pole at the other end with excretory anus ; but the old negative pole (the anus) now becomes positive, becoming a mouth, and a new anus or negative pole is formed at the cut end.[1] *Voilà* transformation of positive into negative.

Another example of polarity in Hæckel : [2] mechanism = monism, and vitalism or teleology=dualism. Already in Kant and Hegel *inner* purpose is a protest against dualism.[3] Mechanism applied to life is a helpless category, at the most we could speak of chemism, if we do not want to renounce all understanding of names. Purpose : Hegel, V, p. 205 : " Mechanism is revealed as a striving of totality even by the fact that it seeks to grasp nature for itself as a whole which requires nothing else for its idea—a totality that is *not to be found in purpose and the extra-mundane understanding connected*

---

[1] This is a rare type of regeneration. A worm more usually behaves like a magnet.
[2] See Appendix II, p. 355.   [3] See Appendix II, p. 356.

with purpose." The point is, however, that mechanism (and also the materialism of the eighteenth century) does not get away from abstract necessity, and hence not from chance either. That matter evolves out of itself the thinking human brain is for him a pure accident, although necessarily determined, step by step, where it happens. But the truth is that it is the nature of matter to advance to the evolution of thinking beings, hence, too, this always necessarily occurs wherever the conditions for it (not necessarily identical at all places and times) are present.

Further, Hegel, V, p. 206 : " Hence this principle (mechanism) gives in its connection of external necessity the consciousness of infinite freedom as against teleology, which puts forward the trivialities and even the despicable aspects of its content as something absolute, in which more general thought can only find itself infinitely cramped and even affected by disgust."

In regard to this, moreover, the colossal waste of matter and motion in nature. In the solar system there are perhaps three planets at most on which life and thinking beings could exist—under present conditions. And the whole enormous apparatus for their sake !

The *inner purpose* in the organism, according to Hegel, V, p. 244,[1] operates through *impulse*. *Pas trop fort*. Impulse is supposed to bring the single living being more or less into harmony with the idea of it. From this it is seen how much the whole *inner purpose* is itself an ideological determination. And yet Lamarck is contained in this.

Valuable self-criticism of the Kantian *thing in itself*, that Kant too suffers shipwrecks on the thinking ego and likewise discovers in it an unknowable thing in itself. Hegel, V, p. 256 *et seq*.

[1] See Appendix II, p. 356.

When Hegel makes the transition from living to knowing by means of propagation [1] (reproduction), there is to be found in this the germ of the theory of evolution, that, organic life once given, it must evolve by the development of the generations to a genus of thinking beings.

The Darwinian theory to be demonstrated as the practical proof of Hegel's account of the inner connection between necessity and chance.[2]

What Hegel calls reciprocal action is the *organic body*, which, therefore, also forms the transition to consciousness, *i.e.* from necessity to freedom, to the idea.   See *Logic*, II, conclusion.

Transformation of quantity into quality.—Simplest example *oxygen* and *ozone*, where 2 : 3 produces quite different properties, even in regard to smell.   Chemistry likewise explains the other allotropic bodies [3] merely by a difference in the number of atoms in the molecule.

If Hegel [4] sees nature as a manifestation of the eternal " idea " in its alienation, and this is such a serious crime, what are we to say of the morphologist Richard Owen : " the archetypal idea was manifested in the flesh under diverse modifications upon this planet, long prior to the existence of those animal species that actually exemplify it " (*Nature of Limbs*, 1849).   If that is said by a mystical natural scientist, who means nothing by it, it is allowed to pass, but if a philosopher says the same thing, and one who means something by it, and in fact *au fond* something correct, although in inverted form, then it is mysticism and a terrible crime.

[1], [2] See Appendix II, p. 357.
[3] *I.e.* different forms of the same substance, for example graphite and diamond.   This explanation is now thought to hold in some cases only.
[4] See Appendix II, p. 358.

The empiricism of observation alone can never adequately prove necessity. *Post hoc* but not *propter hoc* (*Encyclopædia*, I, p. 84). This is so very correct that it does not follow from the continual rising of the sun in the morning that it will rise again to-morrow, and in fact we know now that a time will come when one morning the sun will *not rise*. But the proof of necessity lies in human activity, in experiment, in work : if I am able to make the *post hoc*, it becomes identical with the *propter hoc*.

Chance and Necessity.—Another contradiction in which metaphysics is entangled is that of chance and necessity. What can be more sharply contradictory than these two thought determinations ? How is it possible that both are identical, that the accidental is necessary, and the necessary is also accidental ? Commonsense, and with it the great majority of natural scientists, treats necessity and chance as determinations that exclude one another once for all. A thing, a circumstance, a process is either accidental or necessary, but not both. Hence both exist side by side in nature ; nature contains all sorts of objects and processes, of which some are accidental, the others necessary, and it is only a matter of not confusing the two sorts with one another. Thus, for instance, one assumes the decisive specific characters to be necessary, other differences between individuals of the same species being termed accidental, and this holds good of crystals as it does for plants and animals. Then again the lower group becomes accidental in relation to the higher, so that it is declared to be a matter of chance how many different species are included in the genus *Felis* [1] or *Agnus*, or how many genera and orders there are in a class,[2] and how many individuals of

[1] *E.g. Felis catus*, the cat, *Felis leo*, the lion, *Felis onca*, the jaguar.
[2] This has assumed importance with the work of Willis and others, who find definite laws governing these numbers.

each of these species exist, or how many different species of animals occur in a given region, or what in general the fauna and flora is like. And then it is declared that the necessary is the sole thing of scientific interest and that the accidental is a matter of indifference to science. That is to say : what can be brought under laws, hence what one *knows*, is interesting ; what cannot be brought under laws, and therefore what one does not know, is a matter of indifference and can be ignored. Thereby all science comes to an end, for it has to investigate precisely that which we do *not* know. It means to say : what can be brought under general laws is regarded as necessary, and what cannot be so brought as accidental. Anyone can see that this is the same sort of science as that which proclaims natural what it can explain, and ascribes what it cannot explain to supernatural causes ; whether I term the cause of the inexplicable chance, or whether I term it God, is a matter of complete indifference as far as the thing itself is concerned. Both are only expressions which say : I do not know, and therefore do not belong to science. The latter ceases where the requisite connection is wanting.

In opposition to this view there is determinism, which has passed from French materialism into natural science, and which tries to dispose of chance by denying it altogether. According to this conception only simple, direct necessity prevails in nature. That a particular pea-pod contains five peas and not four or six, that a particular dog's tail is five inches long and not a whit longer or shorter, that this year a particular clover flower was fertilised by a bee and another not, and indeed by precisely one particular bee and at a particular time, that a particular windblown dandelion seed has sprouted and another not, that last night I was bitten by a flea at four o'clock in the morning, and not at three or five o'clock, and on the right shoulder and not on the left calf—these are all facts which have been produced by an irrevocable concatenation of cause and effect, by an unshatterable necessity of such a nature indeed that the gaseous sphere, from which the solar system was derived, was already so constituted that these events had to happen thus and not otherwise. With this kind of necessity we likewise do not get away from the theological conception of nature. Whether with Augustine and Calvin we call it the eternal decree of God, or Kismet as the Turks do, or whether we call it necessity, is all pretty much the same for science. There is no question of tracing the chain of causation in any of these cases ; so we are just as wise in one as in another, the so-called necessity remains an empty phrase, and with it—chance also remains what it was before. As long as we are not able to show on what the number of peas in the pod depends, it remains just a matter of chance, and the assertion that the case was foreseen already in the primordial constitution of the solar system does not get us a step further. Still more. A science which was to set about the task of following back the *casus* of this individual pea-pod in its causal concatenation would be no longer science but pure trifling ; for this same pea-pod alone has in addition innumerable other individual, accidental-seeming qualities : shade of colour, thickness, hardness of the pod, size of the peas, not to speak of the individual peculiarities revealed by the microscope. The one pea-pod, therefore, would already provide more causal connections for following up than all the botanists in the world could solve.

Hence chance is not here explained by necessity, but rather necessity is degraded to the production of what is merely accidental. If the fact that a particular pea-pod contains six peas, and not five or seven, is of the same order as the law of motion of the solar system, or the law of the transformation of energy, then as a matter of fact chance is not elevated into necessity, but rather necessity degraded into chance. Furthermore, however much the diversity of the organic and inorganic species and individuals existing side by side in a given area may be asserted to be based on irrefragable necessity, for the separate species and individuals it remains what it was before, a matter of chance. For the individual animal it is a matter of chance, where it happens to be born, what medium it finds for living, what enemies and how many of them threaten it. For the mother plant it is a matter of chance whither the wind scatters its seeds, and, for the daughter plant, where the seed finds soil for germination; and to assure us that here also everything rests on irrefragable necessity is a poor consolation. The jumbling together of natural objects in a given region, nay more, in the whole world, for all the primordial determination from eternity, remains what it was before—a matter of chance.

In contrast to both conceptions, Hegel came forward with the hitherto quite unheard-of propositions that the accidental has a cause because it is accidental, and just as much also has no cause because it is accidental; that the accidental is necessary, that necessity determines itself as chance, and, on the other hand, this chance is rather absolute necessity (*Logic*, II, Book III, 2 : *Reality*).[1] Natural science has simply ignored these propositions as paradoxical trifling,[2] as self-contra-

---

[1] See Appendix II, p. 358.
[2] Science is now beginning to tackle these questions in connection with quantum mechanics, and will doubtless find a way of expressing them less paradoxically than Hegel's. Meanwhile there seems to be little doubt that many of the laws of ordinary physics are statistical consequences of chance events in atoms. But these chance events are necessary, because, though we cannot predict what a given atom will do, we can predict how many out of a large number will go through a given process.

dictory nonsense, and, as regards theory, has persisted on the one hand in the barrenness of thought of Wolffian metaphysics, according to which a thing is either accidental *or* necessary, but not both at once; or, on the other hand, in the hardly less thoughtless mechanical determinism which by a phrase denies chance in general only to recognise it in practice in each particular case.

While natural science continued to think in this way, what *did it do* in the person of Darwin?

Darwin, in his epoch-making work, set out from the widest existing basis of chance. Precisely the infinite, accidental differences between individuals within a single species, differences which become accentuated until they break through the character of the species, and whose immediate causes even can be demonstrated only in extremely few cases, compelled him to question the previous basis of all regularity in biology, viz. the concept of species in its previous metaphysical rigidity and unchangeability. Without the concept of species, however, all science was nothing. All its branches needed the concept of species as basis: human anatomy and comparative anatomy—embryology, zoology, palæontology, botany, etc., what were they without the concept of species? All their results were not only put in question but directly suspended. Chance overthrows necessity, as conceived hitherto (the material of chance occurrences which had accumulated in the meantime smothered and shattered the old idea of necessity). The previous idea of necessity breaks down. To retain it means dictatorially to impose on nature as a law a human arbitrary determination that is in contradiction to itself and to reality, it means to deny thereby all inner necessity in living nature, it means generally to proclaim the chaotic kingdom of chance to be the sole law of living nature.

" *Gilt nicht mehr der Tausves Jontof,*"[1] cry the biologists of all schools quite logically.

Darwin.

The Struggle for Existence.—Above all this must be
strictly limited to the struggles resulting from plant and
animal *over-population*, which do in fact occur at definite
stages of plant and lower animal life. But one must
keep sharply distinct from it the conditions in which
species alter, old ones die out and newly evolved ones
take their place, *without* this over-population : *e.g.* on
the migration of animals and plants into new regions
where new conditions of climate, soil, etc., are responsible
for the alteration. If *there* the individuals which become
adapted survive and develop into a new species by
continually increasing adaptation, while the other more
stable individuals die away and finally die out, and with
them the imperfect intermediate stages, then this can
and does proceed *without any Malthusianism*, and if the
latter should occur at all it makes no change to the
process, at most it can accelerate it.

Similarly with the gradual alteration of the geographical, climatic, etc., conditions in a given region
(desiccation of central Asia for instance) whether the
members of the animal or plant population there exert
pressure on one another is a matter of indifference ;
the process of evolution of the organisms that is determined by it proceeds all the same. It is the same for
sexual selection, in which case too Malthusianism is quite
unconcerned.

Hence Hæckel's " adaptation and heredity " also can
determine the whole process of evolution, without need
for selection [2] and Malthusianism.

---

[1] "Gone is the authority of the law and the prophets." A line from
one of Heine's poems. See Appendix II, p. 359.

[2] The majority of biologists doubt this to-day.

Darwin's mistake lies precisely in lumping together in " natural selection " or the " survival of the fittest " two absolutely separate things : [1]

1. Selection by the pressure of over-population, where perhaps the strongest survive in the first place, but where the weakest in many respects can also do so.

2. Selection by greater capacity of adaptation to altered circumstances, where the survivors are better suited to these *circumstances*, but where this adaptation as a whole can mean regress just as well as progress (for instance adaptation to parasitic life is *always* regress).

The main thing : that each advance in organic evolution is at the same time a regression, fixing *one-sided* evolution and excluding evolution along many other directions.[2]

This, however, *a basic law.*

[1] *E.g.* the North American rabbits have an eleven-year cycle, in which over-population leads to an epidemic killing most of them. During the year or two of over-population they struggle with one another, during the rest of the cycle there is room for them all.   Haldane has stressed the very different evolutionary effects of these two types of struggle.

[2] *E.g.* the horse has only one toe on each foot, so cannot evolve it into a grasping, climbing, or swimming organ as, for example, the rat could, though the horse is, of course, a better runner than the rat.

Dialectical logic, in contrast to the old, merely formal logic, is not, like the latter, content with enumerating the forms of motion of thought, *i.e.* the various forms of judgement and conclusion, and placing them side by side without any connection. On the contrary, it derives these forms out of one another, it makes one subordinate to another instead of putting them on an equal level, it develops the higher forms out of the lower. Faithful to his division of the whole of logic, Hegel groups judgements as :

1. Judgement of inherence, the simplest form of judgement, in which a general property is affirmatively or negatively predicated of a single thing (positive judgement, the rose is red ; negative, the rose is not blue ; infinite, the rose is not a camel) ;

2. Judgement of subsumption, in which a determination relation is predicated of the subject ; singular judgement : this man is mortal ; particular : some, many men are mortal ; universal : all men are mortal, or man is mortal ;

3. Judgement of necessity, in which its substantial determination is predicated of the subject ; categorical judgement : the rose is a plant ; hypothetical judgement : if the sun rises it is daytime : disjunctive : *Lepidosiren* is either a fish or an amphibian ;

4. Judgement of the notion, in which is predicated of the subject how far it corresponds to its general nature or, as Hegel says, to the notion of it ; *assertoric* judgement : this house is bad ; *problematic* : if a house is constituted in such and such a way, it is good ; *apodeictic* : the house that is constituted in such and such a way is good.

1. *Single Judgement.* 2 and 3. *Special.* 4. *General.*

However dry this sounds here, and however arbitrary at first sight this classification of judgements may here and there appear, yet the inner truth and necessity of this grouping will be illuminating for anyone who studies the brilliant exposition in Hegel's larger *Logic* (Works, V, pp. 63–115). To show how much this grouping is based not only on the laws of thought but also on laws of nature, we would like to put forward here a very well-known example outside this connection.

That friction produces warmth was already known practically to prehistoric man, who discovered the making of fire by friction perhaps more than 100,000 years ago,[1] and who still earlier warmed cold parts of the body by rubbing. But from that to the discovery that friction is in general a source of heat, who knows how many thousands of years elapsed ? Enough that the time came when the human brain was sufficiently developed to be able to formulate the judgement: *friction is a source of heat*, a judgement of inherence, and indeed a positive one.

Still further thousands of years passed until, in 1842, Mayer, Joule, and Colding investigated this special process in its relation to other processes of a similar kind that had been discovered in the meantime, *i.e.* as regards its immediate general conditions, and formulated the judgement : all mechanical motion is capable of being converted into heat by means of friction. So much time and an enormous amount of empirical knowledge were required before we could make the advance in knowledge of the object from the above positive judgement of inherence to this universal judgement of subsumption.

But from now on things went quickly. Only three years later, Mayer was able, at least in substance, to raise the judgement of subsumption to the level at which it now stands.

---

[1] Even *Sinanthropus pekinensis*, who probably lived over 100,000 years ago, and was anatomically very different from modern man, used fire.

Any form of motion, under conditions fixed for each case, is both able and compelled to undergo transformation, directly or indirectly, into any other form of motion : a judgement of the notion, and moreover an apodeictic one, the highest form of judgement altogether.

What, therefore, in Hegel appears as a development of the thought form of judgement as such, confronts us here as the development of our *empirically* based theoretical knowledge of the nature of motion in general. This shows, however, that laws of thought and laws of nature are necessarily in agreement with one another, if only they are correctly known.

We can conceive the first judgement as that of singularity; the isolated fact that friction produces heat is registered. The second judgement is that of particularity : a special form of motion, mechanical motion, exhibits the property, under special conditions (through friction), of passing into another special form of motion, viz. heat. The third judgement is that of universality : any form of motion proves able and compelled to undergo transformation into any other form of motion. In this form the law attains its final expression. By new discoveries we can give new examples of it, we can give it a new and richer content. But we cannot add anything to the law itself as so formulated. In its universality, equally universal in form and content, it is not susceptible of further extension : it is an absolute law of nature.

Unfortunately we are in a difficulty about the form of motion of protein, alias life, so long as we are not able to make protein.[1]

---

[1] We cannot yet make proteins, but we can prepare some of them pure, and, if not alive, they are certainly lively. Thus the pure protein pepsin will break up at least its own weight of another protein per second, and can break up about a million times its weight of protein before wearing out. Others can carry out other similar processes. We are now beginning to study their " form of motion " while doing these things. If we knew enough about their structure to be able to make them, this would, of course, be easier.

Individuality, particularity, universality—these are the three determinations in which the whole " Theory of the Notion " moves.[1] Under these heads, progression takes place not in one but in many modalities, from the single to the particular and from the particular to the universal, and this is often enough exemplified by Hegel as the progression : individual, species, genus. And now the Hæckels come forward with their induction and trumpet it as a great fact—against Hegel—that there is progression from the single to the particular and then to the universal (!), from the individual to the species and then to the genus—and then permit *deductive* conclusions which are supposed to lead further. These people have got into such a deadlock over the opposition between induction and deduction that they reduce all logical forms of conclusion to these two, and in so doing do not notice that (1) they are unconsciously employing quite different forms of conclusion under these names, (2) deprive themselves of the whole wealth of forms of conclusion in so far as it cannot be forced under these two, and (3) thereby convert both forms, induction and deduction, into sheer nonsense.

The above, moreover, demonstrates that judgements involve not Kant's " power of judgement " alone, but at least some power of judgement.

Hofmann (*A Century of Chemistry under the Hohenzollerns*[2]) cites *The Philosophy of Nature*, with a quotation from Rosenkranz, the belletrist whom no real Hegelian recognises. To make the philosophy of nature responsible for Rosenkranz is as foolish as if Hofmann were to make the Hohenzollerns responsible for Marggraf's discovery of beet sugar.

[1], [2] See Appendix II, p. 359.

The eternal laws of nature become transformed more and more into historical ones. That water is fluid from 0°–100° C. is an eternal law of nature, but for it to be valid, there must be (1) water, (2) the given temperature, (3) normal pressure.[1] On the moon there is no water, in the sun only its elements, and the law does not exist for these two heavenly bodies.

The laws of meteorology are also eternal, but only for the earth or for a body of the size, density, axial inclination, and temperature of the earth, and on condition that it has an atmosphere of the same mixture of oxygen and nitrogen and with the same amounts of water vapour being evaporated and precipitated. The moon has no atmosphere, the sun one of glowing metallic vapours; the former has no meteorology, that of the latter is quite different from ours.

Our whole official physics, chemistry, and biology is exclusively *geocentric*, calculated only for the earth. We are still quite ignorant of the conditions of electric and magnetic stress on the sun, fixed stars, and nebulæ, even on the planets of a different density from ours.[2] On the sun, owing to the high temperature, the laws of chemical combination of the elements are suspended or only momentarily operative at the limits of the solar atmosphere, the compounds becoming dissociated again on approaching the sun. The chemistry of the sun, however, is in process of arising, and is necessarily quite different from that of the earth, not overthrowing the latter but standing outside it. In the nebulæ

[1] We could to-day add a fourth condition. The water must be the standard mixture of light and heavy water. For ordinary water is now known to be a mixture of at least six slightly different compounds. No doubt our successors will discover still more conditions.

[2] We now know quite a lot about these matters, thanks to the spectroscope. We know, for example, that many of the atoms in the sun's atmosphere which absorb light are electrically charged, that the sunspots have magnetic fields, and so on.

perhaps there do not exist even those of the 65 elements [1] which are possibly themselves of compound nature. Hence, if we wish to speak of general laws of nature that are uniformly applicable to *all* bodies—from the nebula to man—we are left only with gravity and perhaps the most general form of the theory of the transformation of energy, *vulgo* the mechanical theory of heat.[2] But, on its general logical application to all phenomena of nature, this theory itself becomes converted into a historical presentation of the successive changes occurring in a system of the universe from its origin to its passing away, hence into a history in which at each stage different laws, *i.e.* different phenomenal forms of the same universal motion, predominate, and so nothing remains as continually and universally valid except—*motion*.

Knowing.—Ants have eyes different from ours, they can see the chemical (?) light rays [3] (*Nature*, June 8, 1882, Lubbock), but as regards knowledge of these invisible rays to us, we are considerably more advanced than the ants, and the very fact that we are able to demonstrate *that* ants can see things invisible to us, and that this proof is based solely on perceptions made with *our* eyes, shows that the special construction of the human eye sets no absolute barrier to human cognition.

In addition to the eye, we have not only the other senses but also our thought activity. With regard to the latter again, matters stand exactly as with the eye. To know what can be discovered by our thinking, it is no use, a hundred years after Kant, to try and find out

---

[1] Ninety-two elements (not counting isotopes) are now known. Only a few have yet been detected in gaseous nebulæ, but it is rather doubtful whether many are absent.

[2] We can extend this list now to some laws governing the behaviour of atoms, though even here the gases in the nebulæ emit light according to rather different laws to those on earth, because their atoms are so far apart that they rarely collide.

[3] *I.e.* what we now call ultra-violet radiation. Bees can not merely see it, but distinguish colours within it.

the range of thought from the criticism of the intellect or the investigation of the instrument of knowing. That would be as if Helmholtz were to use the imperfection of our sight (indeed a necessary imperfection, for an eye that could see *all* rays would for that very reason see *nothing at all*),[1] and the construction of the eye, which restricts sight to definite limits and even so does not give quite correct reproduction, as proof that the eye incorrectly or treacherously acquaints us with the nature of what is seen.   What can be ascertained by our thought is more evident from what it has already discovered and is every day still discovering.   And that is already enough both in quantity and quality.   On the other hand, the investigation of the *forms* of thought, the thought determinations, is very profitable and necessary, and since Aristotle this has been systematically undertaken only by Hegel.

In any case we shall never find out *how* chemical rays appear to ants.[2]   Anyone who is worried at this is simply beyond help.

Natural scientists may adopt whatever attitude they please, they will still be under the domination of philosophy.   It is only a question whether they want to be dominated by a bad, fashionable philosophy or by a form of theoretical thought which rests on acquaintance with the history of thought and its achievements.

" Physics, beware of metaphysics," is quite right, but in a contrary sense.

Natural scientists allow philosophy to prolong a pseudo-existence by making shift with the dregs of the

[1] For only certain rays can be brought to a focus by a lens.   If our retina were sensitive to radio waves and X-rays we could not tell from which direction they came, and would at least be greatly handicapped in our vision.

[2] This is perhaps not absolutely certain.   We *may* find that certain physical processes in human brains are always associated with a particular kind of sensation, and identify similar processes in ants.

old metaphysics. Only when natural and historical
science has adopted dialectics will all the philosophical
rubbish—outside the pure theory of thought—be
superfluous, disappearing in positive science.

Hegel, *Geschichte der Philosophie* [*History of Philosophy*]—*Greek philosophy* (*The Ancients' Outlook on
Nature*), I.

Of the first philosophers, Aristotle says (*Metaphysics*,
I, p. 3) that they assert :

" from what all being is, and from what it arises as
the first thing of all, and into what it passes away as
the last thing of all . . . which as substance (οὐσία)
always remains the same and only changes in its
determinations (πάθεσι) this is the element (στοιχεῖον),
and the principle (ἀρχή) of all being. . . . For this
reason they hold that nothing comes into being
(οὔτε γίγνεσθαι οὐδέν) nor passes away, because
the same nature always persists " (p. 198).

Here, therefore, is already the whole original natural
materialism which at its beginning quite naturally
regards the unity of the infinite diversity of natural
phenomena as a matter of course, and seeks it in some
definite corporeal principle, a special thing, as Thales
does in water.

Cicero says : " *Thales* Milesius . . . aquam dixit
esse initium rerum, Deum autem eam mentem, quae
ex aqua cuncta fingeret " (*De Natura Deorum*, I, p. 10).
Hegel quite rightly declares that this is an addition of
Cicero's, and adds : " However, we are not concerned
here with this question whether, in addition, Thales
believed in God ; it is not a matter here of supposition,
belief, popular religion . . . and whether or no he
spoke of God as having created all things from that
water, we would not thereby know anything more of this

being . . . it is an empty word without its idea," p. 209 (*ca.* 600–605).

The oldest Greek philosophers were at the same time investigators of nature : *Thales,* a geometrician, fixed the year at 365 days, and is said to have predicted a solar eclipse. *Anaximander* constructed a sun clock, a kind of map (περίμετρον) of land and sea and various astronomical instruments. *Pythagoras,* a mathematician.

Anaximander of Miletos, according to Plutarch, *Quæst: Convival.,* VIII, p. 8, makes " *man come from a fish, emerging from the water on to the land,*" p. 213. For him the ἀρχὴ καὶ στοιχεῖον τὸ ἄπειρον [1] without determining it as air or water or anything else (διορίζων), Diogenes Lærtius II, paragraph 1. This infinite correctly reproduced by Hegel, p. 215, as " undetermined matter " (*ca.* 580).

Anaximenes of Miletos takes *air* as principle and basic element, declaring it to be infinite (Cicero, *Natura Deorum,* I, p. 10) and that " everything arises from it, in it everything is again dissolved " (Plutarch, *De placitis philos.,* I, p. 3). Here air ἀήρ = πνεῦμα : " Just as our soul, which is air, holds us together, so also a spirit (πνεῦμα) and air hold the whole world together. Spirit and air have the same meaning " (Plutarch). Soul and air conceived as a general medium (*ca.* 555).

Aristotle already says that these more ancient philosophers put the primordial essence in a form of matter : air and water (and perhaps Anaximander in a middle thing between both), later Heraclitus in fire, but none in earth on account of its multiple composition (διὰ τὴν μεγαλομέρειαν). *Metaphysics,* I, 8, p. 217.

Aristotle correctly remarks of all of them that they leave the origin of motion unexplained, p. 218 *et seq.*

Pythagoras of Samos (*ca.* 540) : *number* is the basic

---

[1] " Beginning and element is the infinite."

principle : " That *number* is the essence of all things, and the organisation of the universe as a whole in its determinations is *a harmonious system of numbers and their relations.*" Aristotle, *Metaphysics*, I, 5 *passim.* Hegel justly points out " the audacity of such language, which at one blow strikes down all that is regarded by the imagination as being or as essential (true), and annihilates the sensuous essence," and puts the essence in a thought determination, even if it is a very restricted and one-sided one.   Just as number is subject to definite laws, so also the universe ; hereby its obedience to law was expressed for the first time.   To Pythagoras is ascribed the reduction of musical harmonies to mathematical relations.   Likewise : " The Pythagoreans put fire in the centre, but the earth as a star which revolves in a circle around this central body." Aristotle, *Metaphysics*, I, 5.   This fire, however, is not the sun ; nevertheless this is the first inkling that *the earth moves.*

Hegel on the planetary system :

" . . . the harmonious element, which determines the distances—mathematics has still not been able to give any basis for it.   The empirical numbers are accurately known ; but it has all the appearance of chance, not of necessity.   Any approximate regularity in the distances is known, and thus with luck planets between Mars and Jupiter have been guessed at, where later Ceres, Vesta, Pallas, etc., were discovered ; but astronomy still did not find a logical series in which there was any sense.   Rather it looks with contempt on the regular presentation of this series ; for itself, however, it is an extremely important point which must not be surrendered," p. 267.

For all the naive materialism of the total outlook, the kernel of the later split is already to be found among the most ancient Greeks.   For Thales, the soul is

already something special, something different from the
body (just as he ascribes a soul also to the magnet),
for Anaximenes it is air as in Genesis, for the Pythagoreans it is already immortal and migratory, the body
being purely accidental to it. For the Pythagoreans,
also, the soul is " a chip of the ether (ἀπόσπασμα
αἰθέρος), Diogenes Laertius, VIII, pp. 26-8, where the
cold ether is the air, the dense ether the sea and
moisture.

Aristotle correctly reproaches the Pythagoreans also :
with their numbers "they do not say how motion
comes into being, and how, without motion and change,
there is coming into being and passing away, or states
and activities of heavenly things." *Metaphysics*, I, 8.

Pythagoras is supposed to have known the identity
of the morning and evening star, that the moon gets
its light from the sun, and finally the Pythagorean
theorem.[1] " Pythagoras is said to have slaughtered a
hecatomb on discovering this theorem . . . and however remarkable it may be that his joy went so far on
that account as to order a great feast, to which the rich
and the whole people were invited, it was worth the
trouble. It is joyousness of spirit (knowledge)—at the
expense of the oxen," p. 279.

The Eleatics.

[1] *I.e.* the theorem that the square on the longest side of a right-
angled triangle is the sum of the squares on the shorter sides, *e.g.*
$5^2 = 3^2 + 4^2$.

1. According to Hegel, infinite progress is a barren waste because it appears only as *eternal repetition of the same thing* [1]: $1+1+1$, etc.

2. In reality, however, it is no repetition, but a development, an advance or regression, and thereby it becomes a necessary form of motion. This apart from the fact that it is not infinite : the end of the earth's lifetime can already be foreseen. But then, the earth also is not the whole universe. In Hegel's system, any development was excluded from the temporal history of nature, otherwise nature would not have been the other being of spirit. But in human history the infinite progress of Hegel is recognised as the sole true form of existence of " spirit," except that fantastically this development is assumed to have an end—in the production of the Hegelian philosophy.

3. There is also infinite knowing [2]: *questa infinità che le cose non hanno in progresso, la hanno in giro.* [3] Thus the law of the change of form of motion is an infinite one, including itself in itself. Such infinities, however, are in their turn smitten with finiteness, and only occur sporadically. So also $\frac{1}{r^2}$.

Quantity and Quality.—Number is the purest quantitative determination that we know. But it is chockfull of qualitative differences. 1. Hegel, number and unity, multiplication, division, raising to a higher power, extraction of roots. Thereby, what is not shown in Hegel, qualitative differences already make their appearance : prime numbers and products, simple

[1], [2] See Appendix II, p. 361.
[3] " This infinite, which things do not have in progress, they have in circling."—Galiani, *Della Moneta*, 1803.

roots and powers. 16 is not merely the sum of 16 ones, it is also the square of 4, the fourth power of 2. Still more. Prime numbers communicate new, definitely determined qualities to numbers derived from them by multiplication with other numbers ; only even numbers are divisible by 2, and there is a similar determination in the case of 4 and 8. For 3 there is the rule of the sum of the figures, and the same thing for 9 and also for 6, in the last case in combination with the even number. For 7 there is a special rule. These form the basis for tricks with numbers which seem incomprehensible to the uninitiated. Hence what Hegel says, *quantum*, p. 237,[1] on the absence of thought in arithmetic, is incorrect. Compare, however, " Measure."

As soon as mathematics speaks of the infinitely large and infinitely small, it introduces a qualitative difference which even takes the form of an unbridgeable qualitative opposition. Quantities so enormously different from one another that every rational relation, every comparison, between them ceases, that they become quantitatively incommensurable. The ordinary incommensurability of the circle and the straight line is also a dialectical qualitative difference ; but here it is the difference in *quantity* of *similar magnitudes* that increases the difference of *quality* to the point of incommensurability.

Number.—The single number becomes endowed with quality already in the numerical system itself, and the quality depends on the system used. 9 is not only 1 added together 9 times, but also the basis for 90, 99, 900,000, etc. All numerical laws depend upon and are determined by the system adopted. In dyadic and triadic systems[2] 2 multiplied by 2 does not equal 4, but =

[1] See Appendix II, p. 361.
[2] *I.e.* systems where 2 or 3, not 10, is the so-called radix, so that 100 in the dyadic system means one four plus no two plus no unit, and 11 in the triadic system means one three plus one unit.

100 or =11. In all systems with an odd basic number, the difference between odd and even numbers falls to the ground, e.g. in the system based on 5, 5=10, 10=20, 15=30.[1] Likewise in the same system the sums of digits $3n$ of products of 3 or 9 (6=11, 9=14).[2] Hence the basic number determines not only its own quality but also that of all the other numbers.

With powers of numbers, the matter goes still further : any number can be conceived as the power of any other number--there are as many logarithmic systems as there are whole and fractional numbers.

$\sqrt{-1}$.—The negative magnitudes of algebra are real only in so far as they are connected with positive magnitudes and only within the relation to the latter ; outside this relation, taken by themselves, they are purely imaginary. In trigonometry and analytical geometry, together with the branches of higher mathematics of which these are the basis, they express a definite direction of motion, opposite to the positive direction. But the sine and tangent of the circle can be reckoned from the upper right-hand quadrant just as well as from the right-hand lower quadrant, thus directly reversing plus and minus. Similarly, in analytical geometry, abscissæ [3] can be calculated from the periphery or from the centre of the circle, indeed in all curves they can be reckoned in the direction usually denoted as minus or in any desired direction, and still give a correct

---

[1] In each case the first number of the pair is in the ordinary notation, the second in the system based on 5.

[2] I.e. the rule valid in the ordinary scale of ten that, if a number is divisible by 3 or 9, so is the sum of its digits, does not hold in the scale of 5. (In this system an even number can be detected because the sum of its digits is even. E.g. 10032 in the scale of 5 (642 in the scale of 10) is even because 1+0+0+3+2 is even. A multiple of 3 can be detected because the difference between the sums of alternative digits is zero or a multiple of 3, e.g. 10032 is divisible by 3 because 1+0+2=0+3.)

[3] In modern terminology radii in polar co-ordinates.

rational equation for the curve. Here plus exists only as the complement of minus and *vice versa*. But algebraic abstraction treats it as real and independent, even outside the relation to a *larger*, positive magnitude.

Application of mathematics : in the mechanics of rigid bodies it is absolute, in that of gases approximate,[1] that of fluids already more difficult—in physics more tentative and relative—in chemistry, simple equations of the first order and of the simplest nature—in biology =0.

The differential calculus for the first time made it possible for natural science to represent *processes* mathematically and not only *states* : motion.

That positive and negative can be put as equal, irrespective of which side is positive and which negative : this not only in analytical geometry—still more in physics—see Clausius, p. 87 *et seq.*

Zero, because it is the negation of any definite quantity, is not therefore devoid of content. On the contrary, zero has a very definite content. As the border line between all positive and negative magnitudes, as the sole really neutral number, which can be neither positive nor negative, it is not only a very definite number, but also in itself more important than all other numbers marked off from it. In fact, zero is richer in content than any other number. Put on the right of any other number, it gives in our system of numbers the tenfold value. Instead of zero one could use here any other

[1] This was true when Engels wrote it, but is so no longer. There is an exact mathematics of gases, though the gases treated are abstractions from reality, like rigid bodies. Chemistry is now highly mathematical, and biology moderately so. Even in psychology advanced mathematics are needed in one section, that dealing with tests of intelligence and the like.

sign, but only on the condition that this sign taken by
itself signifies zero=0.   Hence it is part of the nature of
zero itself that it finds this application and that it alone
can be applied.   Zero annihilates every other number
with which it is multiplied ; united with any other
number as divisor or dividend, in the former case it
makes this infinitely large, in the latter infinitely small ;
it is the only number that stands in a relation of infinity
to every other number. $\frac{0}{0}$ can express every number
between $-\infty$ and $+\infty$, and in each case represents a
real magnitude.   The real content of an equation first
clearly emerges when all its members have been brought
on one side, and the equation is thus reduced to zero
value, as already happens for quadratic equations, and
is almost the general rule in higher algebra.   A function,
$F(x, y)=0$ can likewise be put equal to z, and this z,
although it is $=0$, differentiated like an ordinary dependent variable and its partial differential quotient
determined.[1]

The zero of every quantity, however, is itself quantitatively determined, and only on that account is it
possible to calculate with zero.   The very same mathematicians who are quite unembarrassed in reckoning
with zero in the above manner, *i.e.* in operating with it
as a definite quantitative concept, bringing it into
quantitative relation to other quantitative concepts,
clutch their heads in desperation when they read this in
Hegel generalised as : the nothing of a something is a
determinate nothing.

But now for analytical geometry.   Here zero is a
definite point from which measurements are taken along

---

[1] This is done in testing for " double points " on a curve whose
equation is given. *E.g.* if $z=x^3+y^3-3axy=0$ is the equation of a
curve, it crosses itself at the origin, because
$$\frac{dz}{dx}=3x^2-3ay, \quad \frac{dz}{dy}=3y^2-3ax,$$
and both are equal to zero when x and $y=0$.

a line, in one direction positively, in the other negatively. Here, therefore, the zero point has not only just as much significance as any point denoted by a positive or negative magnitude, but a much greater significance than all of them : it is the point on which they are all dependent, to which they are all related, and by which they are all determined. In many cases it can even be taken quite at random. But once adopted, it remains the central point of the whole operation, often determining even the direction of the line along which the other points—the end points of the abscissæ—are to be inserted. If, for example, in order to arrive at the equation of the circle, we choose any point of the periphery as the zero point, then the line of the abscissæ must go through the centre of the circle. All this finds just as much application in mechanics, where likewise the calculation of the motions of the point taken as zero in each case forms the main point and pivot for the entire operation. The zero point of the thermometer is the very definite lower limit of the temperature section that is divided into any desired number of degrees, thereby serving as a measure both for temperature stages within the section as also for higher or lower temperatures. Hence in this case also it is a very essential point. And even the absolute zero of the thermometer in no way represents pure abstract negation, but a very definite state of matter : the limit at which the last trace of independent molecular motion vanishes and matter acts only as mass. Wherever we come upon zero, it represents something very definite, and its practical application in geometry, mechanics, etc., proves that—as limit—it is more important than all the real magnitudes marked off from it.

One.—Nothing looks more simple than quantitative unity, and nothing is more manifold than it, as soon as we investigate it in connection with the corresponding plurality and according to its various modes of origin from plurality. First of all, one is the basic number of the whole positive and negative system of numbers, all other numbers arising by the successive addition of one to itself. One is the expression of all positive, negative, and fractional powers of one : 1^2, $\sqrt{1}$, 1^{-2} are all equal to one. It is the content of all fractions in which the numerator and denominator prove to be equal. It is the expression of every number that is raised to the power of zero, and therewith the sole number the logarithm of which is the same in all systems, viz. $=0$. Thus one is the frontier that divides all possible systems of logarithms into two parts : if the base is greater than one, then the logarithms of all numbers more than one are positive, and of all numbers less than one negative ; if it is smaller than one, the reverse is the case.

Hence, if every number contains unity in itself in as much as it is compounded entirely of ones added together, unity likewise contains all other numbers in itself. This is not only a possibility, in as much as we can construct any number solely of ones, but also a reality, in as much as one is a definite power of every other number. But the very same mathematicians who, without turning a hair, interpolate into their calculations, wherever it suits them, $x° = 1$, or a fraction whose numerator and denominator are equal and, which therefore likewise represents one, who therefore apply mathematically the plurality contained in unity, turn up their noses and grimace if they are told in general terms that unity and plurality are inseparable, mutually penetrating, concepts and that plurality is not less contained in unity than unity is in plurality. How much this is the case we see as soon as we forsake the field of pure numbers. Already in the measurement of lines, surfaces, and the volumes of bodies it becomes apparent that we can take any desired magnitude of the appropriate order as unity, and the same thing holds for measurement of time, weight, motion, etc. For the measurement of cells even millimetres and milligrams are too large, for the measurement of stellar distances or the velocity of light even the kilometre is uncomfortably small, just as the kilogram for planetary and even solar masses. Here is seen very clearly what diversity and multiplicity is contained in the concept of unity, at first sight so simple.

Mathematics.—To commonsense it appears an absurdity to resolve a definite magnitude, *e.g.* a binomial expression, into an infinite series, that is, into something indefinite. But where would we be without infinite series and the binomial theorem ?

Conservation of Energy.—The *quantitative* constancy of motion was already enunciated by Descartes, and indeed almost in the same words as now by ? (Clausius, Robert Mayer ?). On the other hand, the transformation of *form* of motion was only discovered after 1842 and this, not the law of quantitative constancy, is what is new.

Molecule and Differential.—Wiedemann, III, p. 636, puts *finite* and *molecular* distances as directly opposed to one another.[1]

Force and Conservation of Force.—The passages of J. R. Mayer in his two first papers to be cited against Helmholtz.

Trigonometry.—After synthetic geometry has exhausted the properties of a triangle, regarded as such, and has nothing new to say, a more extensive horizon is opened up by a very simple, thoroughly dialectical, procedure. The triangle is no longer considered in and for itself but in connection with another figure, the circle. Every right-angled triangle can be regarded as belonging to a circle : if the hypotenuse $=r$, then the sides enclosing the right angle are sin and cos, if one of these sides $=r$, then the other $=tan$, the hypotenuse $= sec$. In this way the sides and angles are given quite different, definite relationships which without this relation of the triangle to the circle would be impossible to discover and use, and quite a new theory of the triangle arises, far surpassing the old and universally applicable, because every triangle can be resolved into two right-angled triangles. This development of trigonometry from synthetic geometry is a good example

[1] See Appendix II, p. 362.

of dialectics, of the way in which it comprehends things in their connection instead of in isolation.

The consumption of kinetic energy as such within dynamics is always of a twofold nature and has a two-fold result : (1) the kinetic work done, production of a corresponding quantity of potential energy, which, however, is always less than the applied kinetic energy ; (2) overcoming—besides gravity—frictional and other resistances that convert the remainder of the used-up kinetic energy into *heat*. Likewise on re-conversion : according to the way this takes place, a part of the loss through friction, etc., is dissipated as heat—and that is all very ancient !

In the motion of gases—in the process of evaporation—the motion of masses passes directly into molecular motion. Here, therefore, the transition has to be made.

Hegel, *Encyclopædia*, I, p. 205,[1] a prophetic passage on atomic weights in contrast to the physical views of the time, and on atom and molecule as *thought* determinations, on which *thought* has to decide.

Gravity as *the most general determination of materiality* as commonly accepted. That is to say, attraction is a necessary property of matter, but not repulsion. But attraction and repulsion are as inseparable as positive and negative, and hence from dialectics itself it can already be predicted that the true theory of matter must assign as important a place to repulsion as to attraction,[2] and that a theory of matter based

---

[1] See Appendix II, p. 362.

[2] This is emphatically so in modern physics. We can dispense with the notion of attraction by introducing that of the curvature of space-time in the theory of general relativity, and also the notion of inter-change between indistinguishable particles. But that of repulsion remains as a particular case of Pauli's exclusion principle, as part of the very nature of ultimate particles.

on mere attraction is false, inadequate, and one-sided. In fact sufficient phenomena occur that demonstrate this in advance. If only on account of light, the ether is not to be dispensed with. Is the ether of material nature ? It if exists at all, it must be of material nature, it must come under the concept of matter. But it is not affected by gravity.[1] The tail of a comet is granted to be of material nature. It shows a powerful repulsion. Heat in a gas produces repulsion, etc.

Impulse and Friction.—Mechanics regards the effect of an impulse as *purely transitory*. But in reality things are different. On every impact part of the mechanical motion is transformed into heat, and friction is nothing more than a form of impact that continually converts mechanical motion into heat (fire by friction known from primæval times).

Descartes discovered that the ebb and flow of the tides are caused by the attraction of the moon. He also discovered simultaneously with Snell the basic law of the refraction of light, and this in a form peculiar to himself and different from that of Snell.

Theory and Empiricism.—The oblateness of the earth was theoretically established by Newton. The Cassinis and other Frenchmen maintained a long time afterwards, on the basis of their empirical measurements, that the earth is ellipsoidal and the polar axis the longer one.

Aristarchus of Samos, 270 B.C., already held the *Copernican theory of the Earth and Sun*, Mädler, p. 44,[2] Wolf, pp. 35-7.

---

[1] Engels' scepticism as to the reality of the ether has been fully borne out by the development of physics.
[2] See Appendix II, p. 363.

Democritus had already surmised that the *Milky Way* sheds on us the combined light of innumerable small stars, Wolf, p. 313.[1]

A pretty example of the dialectics of nature is the way in which according to present-day theory the *repulsion* of like magnetic Poles is explained by the *attraction* of like electric currents, Guthrie, p. 264.[2]

The contempt of the empiricists for the Greeks receives a peculiar illustration if one reads, for instance, Th. Thomson " On Electricity," where people like Davy and even Faraday grope in the dark (the electric spark, etc.), and arrange experiments that remind one of the stories of Aristotle and Pliny about physico-chemical relations. It is precisely in this new science that the empiricists entirely reproduce the blind groping of the ancients. And when Faraday with his genius gets on the right track, the philistine Thomson has to protest against it (p. 397).

Attraction and Gravitation.—The whole gravitation theory rests on saying that attraction is the essence of matter. This is necessarily false. Where there is attraction, it must be complemented by repulsion. Hence Hegel is quite right in saying that the essence of matter is attraction *and repulsion*. And in fact we are more and more becoming forced to recognise that the dissipation of matter has a limit where attraction is transformed into repulsion, and conversely the condensation of the repelled matter has a limit where it becomes attraction.[3]

[1], [2] See Appendix II, p. 363.
[3] This is startlingly confirmed by modern physics. The spiral nebulæ seem to be flying apart. This is put down by some physicists to a repulsive gravitation at very great distances. Again atomic nuclei which repel one another until they are very close, can cohere to form heavier nuclei if they are brought close enough together. Both these facts were wholly unexpected when Engels wrote.

The earlier, naive conception is as a rule more correct than the later metaphysical one. Thus *Bacon* (and after him Boyle, Newton, and almost all the Englishmen), said heat was motion (Boyle even said molecular motion). It was only in the eighteenth century that the caloric theory arose in France and became more or less accepted on the Continent.

The geocentric standpoint in astronomy is prejudiced and has rightly been abolished. But as we go deeper in our investigations, it comes more and more into its own. . . . The sun, etc., *serve* the earth, Hegel, *Philosophy of Nature*, p. 157. (The whole huge sun exists merely for the sake of the little planets.) Anything other than geocentric physics, chemistry, biology, meteorology, etc., is impossible for us, and it loses nothing by the phrase that this only holds good for the earth and is therefore only relative. If one takes this seriously and demands a centreless science, one puts a stop to *all* science ; [it suffices] us to know that under the same conditions everywhere the same (. . .) [1]

At absolute 0° no gas is possible, all motion of the molecules ceases ; the slightest pressure, and hence their own attraction, forces them together. *Consequently, a permanent gas is an impossibility.* [2]

mv^2 has been proved also for gas molecules by the kinetic theory of gases. Hence there is the same law for molecular motion as for the motion of masses : the difference between the two is here abolished.

(1) Motion in general.
(2) Attraction and repulsion. Transference of motion.

[1] The last line in the manuscript is undecipherable.
[2] Since Engels' death all the gases have proved to be liquefiable by sufficient cold.

(3) Conservation of energy applied to this. Repulsion+attraction—addition of repulsion=energy.

(4) Gravitation—heavenly bodies—terrestrial mechanics.

(5) Physics, heat, electricity.

(6) Chemistry.

(7) Summary.

    (a) Before 4 : mathematically infinite line and + and − are the same.

    (b) In astronomy : performance of work by the tides.

        Double calculation   in Helmholtz, II, p. 120.
        Forces   in Helmholtz, II, p. 190.

Conclusion for Thomson, Clausius, Loschmidt : *The reversion consists in repulsion repelling itself and thereby returning out of the medium into extinct heavenly bodies.* But just therein lies also the proof that repulsion is the really *active* side of motion, and attraction the *passive* side.

(1) Motion of the heavenly bodies. Approximate equilibrium of attraction and repulsion in the motion.

(2) Motion on one heavenly body. Mass. In so far as this from pure mechanical causes, also equilibrium. The masses at rest on their foundation. On the moon this equilibrium is apparently complete. Mechanical attraction has overcome mechanical repulsion. From the standpoint of pure mechanics, we do not know what has become of the repulsion, and pure mechanics just as little explains whence come the " forces," by which, for example, masses on the earth move *against* gravity. It takes the fact for granted. Hence here there is simple communication of repelling, displacing motion from mass to mass, with equality of attraction and repulsion.

(3) The overwhelming majority of all terrestrial motions, however, are made up of the conversion of one form of motion into another, mechanical motion into heat, electricity, and chemical motion, and one into another, hence either the transformation of attraction into repulsion (mechanical motion into heat, electricity, chemical decomposition) (the transformation is the conversion of the original, *lifting* mechanical motion into heat, not of the *falling motion*, which is only the semblance).

(4) All energy now active on the earth is transformed heat from the sun.[1]

How little Comte can have been the author of his encyclopædic arrangement of the natural sciences, which he copied from St. Simon, is already evident from the fact that it only serves him for the purpose of *arranging the means of instruction* and *course of instruction*, and so leads to the crazy *enseignement intégral*, where one science is always exhausted before another is even broached, pushing a basically correct idea to a mathematical absurdity.

Physiography.—After the transition from chemistry to life has been made, then in the first place it is necessary to analyse the conditions in which life has been produced and continues to exist, *i.e.* first of all geology, meteorology, and so on. Then the various forms of life themselves, which indeed without this are incomprehensible.

A new epoch begins in chemistry with atomistics (hence Dalton, not Lavoisier, is the father of modern chemistry), and correspondingly in physics with the molecular theory (in a different form, but essentially

---

[1] This is nearly but not quite true. The energy of the tides is transformed relative motion of the earth and moon. That of volcanoes is in part derived from radio-activity.

representing only the other side of this process, with the discovery of the transformation of the forms of motion). The new atomistics distinguishes itself from all previous to it by the fact that it does not maintain (idiots excepted) that matter is *merely* discrete, but that the discrete parts at various stages (ether atoms, chemical atoms, masses, stellar bodies) are various *nodal points* [1] which determine the various *qualitative* modes of existence of matter in general—down to weightlessness and repulsion.

Hegel constructed his theory of light and colour out of pure thought, and in so doing fell into the *grossest empiricism* of homebred philistine experience (although with a certain justification, since this point had not been cleared up at that time), *e.g.* where he adduced against Newton the mixtures of colours used by painters, p. 314, below.

Static and Dynamic Electricity.—Static or frictional electricity is the subjection to stress of the electricity occurring in nature, already existing in the *form* of electricity but in an equilibrated, neutral state. Hence the removal of this stress—if and in so far as the electricity can be propagated by conduction—also occurs at one stroke, by a spark, which re-establishes the neutral state.

Dynamic or voltaic electricity, on the other hand, is electricity produced by the conversion of chemical motion into electricity. Under certain definite conditions, it is produced by the solution of zinc, copper, etc. Here the stress is not acute, but chronic. At every moment new positive and negative electricity is pro-

---

[1] This again has been fully borne out by modern developments. The atom is a unit for the purposes of ordinary chemistry, but if we employ forces on an altogether different scale of magnitude from those used in chemistry, we can break up atoms or put them together.

duced from another form of motion, and not already existing $\pm$ electricity separated into $+$ and $-$. The process is a continuous one, and thus too its result, the electricity, does not take the form of instantaneous stress and discharge, but of a continuous current which can be reconverted at the poles into the chemical motion from which it arose. This process is termed electrolysis. In this process, as well as in the production of electricity by chemical decomposition (in which electricity is liberated instead of heat, and in fact as much electricity as under other circumstances is set free as heat, Grove,[1] p. 210), the current can be traced in the liquid (exchange of atoms in adjacent molecules—this is the current).[2]

This electricity, being of the nature of a current, for that very reason cannot be directly converted into static electricity. By means of induction, however, neutral electricity already existing as such can be de-neutralised. In the nature of things the induced electricity has to follow that which induces it, and therefore must likewise be of a flowing character. On the other hand, this obviously gives the possibility of condensing the current and of converting it into static electricity, or rather into a higher form that combines the property of a current with that of stress. This is solved by Ruhmkorff's machine. It provides an inductional electricity, which achieves this result.

While Coulomb speaks of "*particles* of electricity, which repel each other inversely as the square of their distance," Thomson calmly takes this as proved, p. 358.[3] Ditto, p. 366,[3] the hypothesis that electricity consists of two fluids, positive and negative, whose particles repel each other. That electricity in a charged body is retained merely by the pressure of the atmosphere,

---

[1] See Appendix II, p. 364.
[2] This sentence would have to be considerably revised in view of recent developments.   [3] See Appendix II, p. 364.

p. 360.[1] Faraday put the seat of electricity in the opposed poles of the atoms (or molecules, they are still much mixed up), and thus for the first time stated that electricity is not a fluid but a form of motion, a " force," p. 378.[1] What old Thomson cannot get into his head at all is that it is precisely the spark that is of a *material* nature !

Already, in 1822, Faraday discovered that the momentary induced current—the first as well as the second reversed current—" participates more of the current produced by the discharge of the Leyden jar than that produced by the voltaic battery "—herein lay the whole secret, p. 385.

The spark has been the subject of all sorts of cock and bull stories, which are now known to be special cases or illusions : the spark from a positive body is said to be a " pencil of rays, brush, or cone," the point of which is the point of discharge; the negative spark, on the other hand, is said to be a " *star*," p. 369. A short spark is said to be always white, a long one usually reddish or violet coloured (wonderful nonsense of Faraday on the spark, p. 400).[1] The spark, drawn from the prime conductor by a metal sphere is said to be white, by the hand—purple, by aqueous moisture—red (p. 405).[1] The spark, *i.e.* light, is said to be " not inherent in electricity but merely the result of the compression of the air. That air is violently and suddenly *compressed* when an electric spark pushes through it " is proved by the experiment of Kinnersley in Philadelphia, according to which the spark produces " a sudden *rarefaction of the air in the tube*," and drives the water into the tube, p. 407. In Germany, 30 years ago, Winterl and others believed that the spark, or electric light, was of the same nature as fire and arises by the union of two electricities. Against which Thomson

[1] See Appendix II, pp. 365-366.

seriously proves that the point where the two electricities meet is precisely the poorest in light, and that
it is two-thirds from the positive and one-third from the
negative end ! (pp. 409–10).[1] That fire is here still
something quite *mythical* is obvious.

With the same seriousness is reported the experiment
of Dessaignes, according to which, with a rising barometer and falling temperature, glass, resin, silk, etc.,
become negatively electrified on being plunged into
mercury, but positively electrified if the barometer is
falling and the temperature rising, and in summer always
become positive in impure, and always negative in pure,
mercury, that in summer gold and various other metals
become positive on warming and negative on cooling,
the reverse being the case in winter, that they are highly
electric with a high barometer and northerly wind,
positive if the temperature is rising, negative if falling,
etc., p. 416.[2]

How matters stood in regard to *heat* : " in order to
produce thermo-electric effects, it is not necessary to
apply heat. Anything *which alters the temperature* in
one part of the chain also occasions a deviation in the
declination of the magnet." For instance, the cooling
of a metal by ice or evaporation of ether ! p. 419.[3]

The electro-chemical theory, p. 438,[4] accepted as " at
least very ingenious and plausible."

Fabroni and Wollaston had already long ago, and
Faraday recently, asserted that voltaic electricity was
the simple consequence of chemical processes, and
Faraday had even given the correct explanation of the
shifting of atoms taking place in the liquid, and established
that the quantity of electricity was to be measured by
the quantity of the electrolytic product.

With the help of Faraday he arrives at the law : " that

[1] See Appendix II, p. 366.   [2] See Appendix II, p. 367.
[3], [4] See Appendix II, p. 368.

every atom must be naturally surrounded by the same quantity of electricity, so that in this respect heat and electricity resemble each other " !

Electricity.—In regard to Thomson's cock and bull stories, *cf.* Hegel, pp. 346–7, where there is exactly the same thing. On the other hand, Hegel already conceives frictional electricity quite clearly as *stress*, in contrast to the fluid theory and the electrical matter theory, p. 347.

Hegel's division (the original one) into mechanics, chemics, and organics, fully adequate for the time. Mechanics : the movement of masses. Chemistry : molecular motion (for physics is also included in this and, indeed, both belong to the same order) and atomic motion. Organics : the motion of bodies in which the two are inseparable. For the organism is certainly *the higher unity which within itself unites mechanics, physics, and chemistry into a whole* where the trinity can no longer be separated. In the organism, mechanical motion is effected directly by physical and chemical change, in the form of nutrition, respiration, secretion, etc., just as much as pure muscular movement.

Each group in turn is two-fold. Mechanics : (1) stellar, (2) terrestrial. Molecular motion : (1) physics, (2) chemistry. Organics : (1) plants, (2) animals.

Electrochemistry.—In describing the effect of the electric spark in chemical decomposition and synthesis, Wiedemann [1] declares that this is more the concern of chemistry. In the same case the chemists declare that it is rather a matter which concerns physics. Thus at the point of contact of molecular and atomic

[1] See Appendix II, p. 368.

science, both declare themselves incompetent, while it is precisely *at this point that the biggest results are to be expected*.[1]

How ancient, convenient methods, adapted to previously customary practice, become transferred to other branches and there are a hindrance : in chemistry, the calculation of composition in percentages, which was the most suitable method of all for making it impossible to discover the laws of constant proportion and multiple proportion in combination, and indeed did make them undiscoverable for long enough.[2]

[1] This is an example of the extreme power of the dialectical method. It was just the study of electrically charged atoms and molecules which led to the discovery of the electron and of atomic structure.

[2] *E.g.* the relation between carbon monoxide and carbon dioxide is obscure when we say that the first contains 42·9 per cent. carbon and 57·1 per cent. oxygen, the second 27·3 per cent. carbon, and 72·7 per cent. oxygen. It is clear when we say that the first contains 1 part of carbon to 1·33 of oxygen, and the second 1 of carbon to 2·67 of oxygen.

(1) Historical introduction : the metaphysical out-look has become impossible in natural science owing to the very development of the latter.

(2) Course of the theoretical development in Germany since Hegel (old preface). The return to dialectics takes place unconsciously, hence contradictory and slow.

(3) Dialectics as the science of the total connections. Main laws : transformations of quantity and quality—mutual penetration of polar opposites and transformation into each other when driven to extremes—development by contradiction or negation of the negation—spiral form of development.

(4) The inter-connection of the sciences. Mathematics, mechanics, physics, chemistry, biology (Comte) St. Simon, and Hegel.

(5) Surveys of the separate sciences and their dialectical content :

1. Mathematics : dialectical aids and expressions—mathematical infinite really occurring.

2. Celestial mechanics—now merged into a *process*.
   —Mechanics : point of departure was inertia, which is only the negative expression of the indestructibility of motion.

3. Physics—passage of the molecular motions into one another. Clausius and Loschmidt.

4. Chemistry : theories, energy.

5. Biology. Darwinism. Necessity and chance.

6. The limits of knowledge. Dubois-Reymond and Nägeli—Helmholtz, Kant, Hume.

7. The mechanistic theory—Hæckel.

8. The plastidule [1] soul—Hæckel and Nägeli.

[1] The plastidule was a primitive living unit smaller than the cell postulated by Hæckel, on rather inadequate grounds, more or less anticipating the gene. It was supposed to have a soul.

9. Science and teaching [1]—Virchow.

10. The cell state—Virchow.

11. Darwinian politics and theory of society—Hæckel and Schmidt. Differentiation of human beings through *labour*. Application of economics to natural science. Helmholtz's "*Work*" (Popular Lectures II).

---

[1] Engels refers here to the pamphlet by Virchow, *Die Freiheit der Wissenschaft im modernen Staat* [*The Freedom of Science in the Modern State*], published in Berlin, 1877, and Hæckel's reply, *Freie Wissenschaft und freie Lehre* [*Free Science and Free Teaching*].

VIII

TIDAL FRICTION, KANT AND THOMSON-TAIT ON THE ROTATION OF THE EARTH AND LUNAR ATTRACTION.

THOMSON and Tait, *Nat. Philos.*, I, p. 191 (paragraph 276) :

"There are also indirect resistances, owing to friction impeding the tidal motions, on all bodies which, like the earth, have portions of their free surfaces covered by liquid, which, as long as these bodies move relatively to neighbouring bodies, must keep drawing off energy from their relative motions. Thus, if we consider, in the first place, the action of the moon alone, on the earth with its oceans, lakes, and rivers, we perceive that it must tend to equalise the periods of the earth's rotation about its axis, and of the revolution of the two bodies about their centre of inertia ; because as long as these periods differ, the tidal action of the earth's surface must keep subtracting energy from their motions. To view the subject more in detail, and, at the same time, to avoid unnecessary complications, let us suppose the moon to be a uniform spherical body, the mutual action and reaction of gravitation between her mass and the earth's will be equivalent to a single force in some line through her centre ; *and must be such as to impede the earth's rotation as long as this is performed in a shorter period than the moon's motion round the earth.* It must, therefore, lie in some such direction as the line MQ in the diagram, which represents, necessarily with enormous exaggeration, its deviation, OQ, from the earth's centre. Now the actual force on the moon in the line MQ may be regarded as consisting of a force in the line MO towards the earth's centre, sensibly

271

equal in amount to the whole force, and a comparatively very small force in the line MT perpendicular
to MO. This latter is very nearly tangential to the
moon's path, and is in the direction *with* her motion.
Such a force, if suddenly commencing to act, would,
in the first place, increase the moon's velocity; but
after a certain time she would have moved so much
farther from the earth, in virtue of this acceleration,
as to have lost, by moving against the earth's attrac-

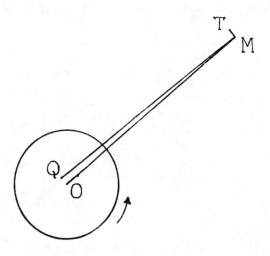

tion, as much velocity as she had gained by the
tangential accelerating force. The effect of a continued tangential force, acting with the motion, but
so small in amount as to make only a small deviation
at any moment from the circular form of the orbit, is
to gradually increase the distance from the central
body, and to cause as much again as its own amount of
work to be done against the attraction of the central
mass, by the kinetic energy of motion lost. The
circumstances will be readily understood by considering this motion round the central body in a very
gradual spiral path tending outwards. Provided the

law of force is the inverse square of the distance, the tangential component of gravity against the motion will be twice as great as the disturbing tangential force in the direction with the motion ; and therefore one-half of the amount of work done against the former, is done by the latter, and the other half by kinetic energy taken from the motion. The integral effect on the moon's motion, of the particular disturbing cause now under consideration, is most easily found by using the principle of moments of momenta. Thus we see that as much moment of momentum is gained in any time by the motions of the centres of inertia, of the moon and earth relatively to their common centre of inertia, as is lost by the earth's rotation about its axis. The sum of the moments of momentum of the centres of inertia of the moon and earth as moving at present, is about 4·45 times the present moment of momentum of the earth's rotation.

The average plane of the former is the ecliptic ; and therefore the axes of the two moments are inclined to one another at the average angle of 23° 27·5′, which, as we are neglecting the sun's influence on the plane of the moon's motion, may be taken as the actual inclination of the two axes at present. The resultant, or whole moment of momentum, is therefore 5·38 times that of the earth's present rotation, and its axis is inclined 19° 13′ to the axis of the earth. Hence the ultimate tendency of the *tides* is to reduce the earth and moon to a simple uniform rotation with this resultant moment round this resultant axis, as if they were two parts of one rigid body : in which condition the moon's distance would be increased (approximately) in the ratio 1 : 1·46, being the ratio of the square of the present moment of momentum of the centres of inertia to the square of the whole moment of momentum ; and the period of revolution in the ratio 1 : 1·77, being that of the cubes of the same quantities. The distance would therefore be increased to 347,100 miles, and the period lengthened to 48·36 days. Were there no other body in the

universe but the earth and the moon, these two bodies
might go on moving thus for ever, in circular orbits
round their common centre of inertia, and the earth
rotating about its axis in the same period, so as always
to turn the same face to the moon, and, therefore, to
have all the liquids at its surface at rest relatively
to the solid.   But the existence of the sun would
prevent any such state of things from being permanent.
There would be solar tides—twice high water and
twice low water—in the period of the earth's revolution relatively to the sun (that is to say, twice in the
solar day, or, which would be the same thing, the
month).   This could not go on without *loss of energy
by fluid friction.*   It is not easy to trace the whole
course of the disturbance in the earth's and moon's
motions which this cause would produce, but its
ultimate effect must be to bring the earth, moon, and
sun to rotate round their common centre of inertia,
like parts of one rigid body." [1]

Kant, in 1754, was the first to put forward the view
that the rotation of the earth is retarded by tidal friction
and that this effect will only reach its conclusion " when
its (the earth's) surface will be at relative rest in relation
to the moon, *i.e.* when it will rotate on its axis in the
same period that the moon takes to revolve round the
earth, and consequently will always turn the same side
to the latter."   He held the view that this retardation
had its origin in tidal friction alone, arising, therefore,
from the presence of fluid masses on the earth :

" If the earth were a quite solid mass without any
fluid, neither the attraction of the sun nor of the
moon would do anything to alter its free axial rotation ; for it draws with equal force both the eastern
and western parts of the terrestrial sphere and so does
not cause any inclination either to the one or to the

[1] This theory has since been greatly developed, and the actual rate
at which tidal friction is lengthening the day has been approximately
found.

other side; consequently it allows the earth full freedom to continue this rotation unhindered as if there were no external influence on it."

Kant could rest content with this result. All scientific pre-requisites were lacking at that time for penetrating deeper into the effect of the moon on the rotation of the earth. Indeed, it required almost a hundred years before Kant's theory obtained general recognition, and still longer before it was discovered that the ebb and flow of the tides are only the *visible* aspect of the effect exercised by the attraction of the sun and moon on the rotation of the earth.

This more general conception of the matter is just that which has been developed by Thomson and Tait. The attraction of the moon and sun affects not only the fluids of the terrestrial body or its surface, but the whole mass of the earth in general in a manner that hinders the rotation of the earth. As long as the period of the earth's rotation does not coincide with the period of the moon's revolution round the earth, so long the attraction of the moon—to deal with this alone first of all—has the effect of bringing the two periods closer and closer together. If the rotational period of the (relative) central body were longer than the period of revolution of the satellite, the former would be gradually lengthened ;[1] if it were shorter, as is the case for the earth, it would be slowed down. But neither in the one case will kinetic energy be created out of nothing, nor in the other will it be annihilated. In the first case, the satellite would approach closer to the central body and shorten its period of revolution, in the second it would increase its distance from it and acquire a longer period of revolution. In the first case, the satellite by approaching the central body loses exactly as much

---

[1] A slip of the pen ; the word should obviously be "shortened."

potential energy as the central body gains in kinetic energy from the accelerated rotation; in the second case the satellite, by increasing its distance gains exactly the same amount of potential energy as the central body loses in kinetic energy of rotation. The total amount of dynamic energy, potential and kinetic, present in the earth-moon system remains the same; the system is fully conservative.[1]

It is seen that this theory is entirely independent of the physico-chemical constitution of the bodies concerned. It is derived from the general laws of motion of free heavenly bodies, the connection between them being produced by attraction in proportion to their masses and inverse proportion to the square of the distances between them. The theory has obviously arisen as a generalisation of Kant's theory of tidal friction, and is even presented here by Thomson and Tait as its substantiation on mathematical lines. But in reality—and remarkably enough the authors have simply no inkling of this—in reality it excludes the special case of tidal friction.

Friction is hindrance to the motion of mass, and for centuries it was regarded as the destruction of such motion, and therefore of kinetic energy. We now know

---

[1] There can be no doubt that Engels was right when he pointed out Thomson and Tait's error in saying that the changes in the length of the day and month "could not go on without loss of energy by fluid friction." We now know that there are tides in the earth as well as in the ocean. But Engels was wrong in supposing that the moon could move away from the earth without loss of energy. For in a system such as the earth and moon the angular momentum (moment of momentum) remains constant unless it is diminished or increased by the tidal action of some external body. If both momentum and energy are conserved no systematic slowing down can occur. This is readily seen in the simplified case where the moon is supposed to go round in a circle in the plane of the earth's equator. In this case there are only two possible variables, the lengths of the day and month. But so long as the moment of momentum and the energy of the system are unchanged we have two equations to determine these quantities, and they are therefore fixed.

that friction and impact are the two forms in which kinetic energy is converted into molecular energy, into heat. In all friction, therefore, kinetic energy as such is lost in order to re-appear, not as potential energy in the sense of dynamics, but as molecular motion in the definite form of heat. The kinetic energy lost by friction is, therefore, in the first place *really lost* for the dynamic aspects of the system concerned. It can only become dynamically effective again if it is *re-converted* from the form of heat into kinetic energy.

How then does the matter stand in the case of tidal friction ? It is obvious that here also the whole of the kinetic energy communicated to the masses of water on the earth's surface by lunar attraction is converted into heat, whether by friction of the water particles among themselves in virtue of the viscosity of the water, or by friction at the rigid surface of the earth and the comminution of rocks which stand up against the tidal motion. Of this heat there is re-converted into kinetic energy only the infinitesimally small part that contributes to evaporation at the surface of the water. But even this infinitesimally small amount of kinetic energy, leaving the total system earth-moon at a part of the earth's surface, remains first of all subject to the conditions prevailing at the earth's surface, and these conditions lead to all energy active there reaching one and the same final destiny : final conversion into heat and radiation into space.

Consequently, to the extent that tidal friction indisputably acts in an impeding manner on the rotation of the earth, the kinetic energy used for this purpose is absolutely lost to the dynamic system earth-moon. It can therefore not re-appear within this system as dynamic potential energy. In other words, of the kinetic energy expended in impeding the earth's rotation by means of the attraction of the moon, only that part that acts on the *solid* mass of the earth's body can entirely re-appear as dynamic potential energy, and hence be compensated for by a corresponding increase of the distance of the moon.   On the other hand, the part that acts on the fluid masses of the earth can do so only in so far as it does not set these masses themselves into a motion opposite in direction to that of the earth's rotation, for such a motion is *wholly* converted into heat and is finally lost to the system by radiation.

What holds good for tidal friction at the surface of the earth is equally valid for the so often hypothetically assumed tidal friction of a supposed fluid nucleus of the earth's interior.

The most peculiar part of the matter is that Thomson and Tait do not notice that in order to establish the theory of tidal friction they are putting forward a theory that proceeds from the tacit assumption that the earth is *an entirely rigid body*,[1] and so exclude any possibility of tidal flow and hence also of tidal friction.

[1] Although Engels formulated his criticism of Thomson and Tait incorrectly, he was right in a fundamental point.   The earth-moon system would evolve in such a way as to lengthen the day and month even if there were no ocean.   For the earth is a *solid* (*fester*) body, but not a *rigid* (*starrer*) body in the sense in which this latter word is used in theoretical mechanics, that is to say a body whose shape is unaltered by the forces on it.   Of course a rigid body is a mathematical abstraction, like a flat surface.   There are no perfectly rigid bodies nor flat surfaces.   And it has now been shown that the solid earth bends slightly as the moon's attraction varies.   There are solid tides as well as liquid tides though much smaller.   These act in the same way as the tides in the ocean, though much more slowly.

IX

THE PART PLAYED BY LABOUR IN THE TRANSITION FROM APE TO MAN

LABOUR is the source of all wealth, the economists assert. It is this—next to nature, which supplies it with the material that it converts into wealth. But it is also infinitely more than this. It is the primary basic condition for all human existence, and this to such an extent that, in a sense, we have to say that labour created man himself.

Many hundreds of thousands of years ago, during an epoch, not yet definitely determined, of that period of the earth's history which geologists call the Tertiary period, most likely towards the end of it, a specially highly-developed race of anthropoid apes lived somewhere in the tropical zone—probably on a great continent that has now sunk to the bottom of the Indian Ocean.[1] Darwin has given us an approximate description of these ancestors of ours. They were completely covered with hair, they had beards and pointed ears, and they lived in bands in the trees.

Almost certainly as an immediate consequence of their mode of life, for in climbing the hands fulfil quite different functions from the feet, these apes when moving on level ground began to drop the habit of using their hands and to adopt a more and more erect posture in walking. This was *the decisive step in the transition from ape to man.*

---

[1] This is rather unlikely. A broad ridge across the Indian Ocean has been found in the region indicated, but if it represents a sunken continent, this probably sank before our ancestors had evolved so far.

All anthropoid apes of the present day can stand erect and move about on their feet alone, but only in case of need and in a very clumsy way. Their natural gait is in a half-erect posture and includes the use of the hands. The majority rest the knuckles of the fist on the ground and, with legs drawn up, swing the body through their long arms, much as a cripple moves with the aid of crutches. In general, we can to-day still observe among apes all the transition stages from walking on all fours to walking on two legs. But for none of them has the latter method become more than a makeshift.

For erect gait among our hairy ancestors to have become first the rule and in time a necessity presupposes that in the meantime the hands became more and more devoted to other functions.[1] Even among the apes there already prevails a certain separation in the employment of the hands and feet. As already mentioned, in climbing the hands are used differently from the feet. The former serve primarily for collecting and holding food, as already occurs in the use of the fore paws among lower mammals. Many monkeys use their hands to build nests for themselves in the trees or even, like the chimpanzee, to construct roofs between the branches for protection against the weather. With their hands they seize hold of clubs to defend themselves against enemies, or bombard the latter with fruits and stones. In captivity, they carry out with their hands a number of simple operations copied from human beings.[2] But it is just here that one sees how great is the gulf between the undeveloped hand of even the most anthropoid of apes and the human hand that has been highly perfected by the labour of hundreds of thousands of years. The number and general arrange-

[1] It has been suggested that this process was speeded up by the dying out of forests in central Asia, so that our ancestors were forced to run after their prey.
[2] Chimpanzees can carry out some operations on their own initiative.

ment of the bones and muscles are the same in both ;
but the hand of the lowest savage can perform hundreds
of operations that no monkey's hand can imitate.
No simian hand has ever fashioned even the crudest
stone knife.

At first, therefore, the operations, for which our
ancestors gradually learned to adapt their hands during
the many thousands of years of transition from ape
to man, could only have been very simple. The lowest
savages, even those in whom a regression to a more
animal-like condition, with a simultaneous physical
degeneration, can be assumed to have occurred, are
nevertheless far superior to these transitional beings.
Before the first flint could be fashioned into a knife by
human hands, a period of time must probably have
elapsed in comparison with which the historical period
known to us appears insignificant. But the decisive
step was taken : *the hand became free* and could henceforth attain ever greater dexterity and skill, and the
greater flexibility thus acquired was inherited and
increased from generation to generation.

Thus the hand is not only the organ of labour, *it is also
the product of labour.* Only by labour, by adaptation to
ever new operations, by inheritance of the resulting
special development of muscles, ligaments, and, over
longer periods of time, bones as well, and by the ever-
renewed employment of these inherited improvements
in new, more and more complicated operations, has the
human hand attained the high degree of perfection
that has enabled it to conjure into being the pictures of
Raphael, the statues of Thorwaldsen, the music of
Paganini.

But the hand did not exist by itself. It was only one
member of an entire, highly complex organism. And
what benefited the hand, benefited also the whole
body it served ; and this in two ways.

In the first place, the body benefited in consequence of the law of correlation of growth, as Darwin called it. According to this law, particular forms of the individual parts of an organic being are always bound up with certain forms of other parts that apparently have no connection with the first. Thus all animals that have red blood cells without a cell nucleus, and in which the neck is connected to the first vertebra by means of a double articulation (condyles), also without exception possess lacteal glands for suckling their young. Similarly cloven hooves in mammals are regularly associated with the possession of a multiple stomach for rumination. Changes in certain forms involve changes in the form of other parts of the body, although we cannot explain this connection.[1] Perfectly white cats with blue eyes are always, or almost always, deaf. The gradual perfecting of the human hand, and the development that keeps pace with it in the adaptation of the feet for erect gait, has undoubtedly also, by virtue of such correlation, reacted on other parts of the organism. However, this action has as yet been much too little investigated for us to be able to do more here than to state the fact in general terms.

Much more important is the direct, demonstrable reaction of the development of the hand on the rest of the organism. As already said, our simian ancestors were gregarious ; it is obviously impossible to seek the derivation of man, the most social of all animals, from non-gregarious immediate ancestors. The mastery over nature, which begins with the development of the hand, with labour, widened man's horizon at every new advance. He was continually discovering new, hitherto unknown, properties of natural objects. On the other

---

[1] The connection can now be explained in a few cases. Thus white onions are more susceptible to moulds than the coloured forms, because they lack an antiseptic substance as well as colouring matter. The antiseptic is a necessary stage in building up the pigment.

hand, the development of labour necessarily helped to bring the members of society closer together by multiplying cases of mutual support, joint activity, and by making clear the advantage of this joint activity to each individual. In short, men in the making arrived at the point where *they had something to say* to one another. The need led to the creation of its organ ; the undeveloped larynx of the ape was slowly but surely transformed by means of gradually increased modulation, and the organs of the mouth gradually learned to pronounce one articulate letter after another.

Comparison with animals proves that this explanation of the origin of language from and in the process of labour is the only correct one. The little that even the most highly-developed animals need to communicate to one another can be communicated even without the aid of articulate speech. In a state of nature, no animal feels its inability to speak or to understand human speech. It is quite different when it has been tamed by man. The dog and the horse, by association with man, have developed such a good ear for articulate speech that they easily learn to understand any language within the range of their circle of ideas. Moreover they have acquired the capacity for feelings, such as affection for man, gratitude, etc., which were previously foreign to them. Anyone who has had much to do with such animals will hardly be able to escape the conviction that there are plenty of cases where they *now* feel their inability to speak is a defect, although, unfortunately, it can no longer be remedied owing to their vocal organs being specialised in a definite direction. However, where the organ exists, within certain limits even this inability disappears. The buccal organs of birds are of course radically different from those of man, yet birds are the only animals that can learn to speak ; and it is the bird with the most hideous voice, the parrot, that speaks best of all.   It need not be objected that the parrot does not understand what it says.   It is true that for the sheer pleasure of talking and associating with human beings, the parrot will chatter for hours at a time, continuing to repeat its whole vocabulary. But within the limits of its circle of ideas it can also learn to understand what it is saying.   Teach a parrot swear words in such a way that it gets an idea of their significance (one of the great amusements of sailors returning from the tropics);   on teasing it one will soon discover that it knows how to use its swear words just as correctly as a Berlin costermonger.   Similarly with begging for titbits.

First comes labour, after it, and then side by side with it, articulate speech—these were the two most essential stimuli under the influence of which the brain of the ape gradually changed into that of man, which for all its similarity to the former is far larger and more perfect.   Hand in hand with the development of the brain went the development of its most immediate instruments—the sense organs.   Just as the gradual development of speech is inevitably accompanied by a corresponding refinement of the organ of hearing, so the development of the brain as a whole is accompanied by a refinement of all the senses.   The eagle sees much farther than man, but the human eye sees considerably more in things than does the eye of the eagle.   The dog has a far keener sense of smell than man, but it does not distinguish a hundredth part of the odours that for man are definite features of different things.[1]   And the sense of touch, which the ape hardly possesses in its crudest initial form, has been developed side by side with the development of the human hand itself, through the medium of labour.

----

[1] This is doubtful. A dog cannot distinguish between smells which are distinct to men, but the converse is also true.

The reaction on labour and speech of the development of the brain and its attendant senses, of the increasing clarity of consciousness, power of abstraction and of judgement, gave an ever-renewed impulse to the further development of both labour and speech. This further development did not reach its conclusion when man finally became distinct from the monkey, but, on the whole, continued to make powerful progress, varying in degree and direction among different peoples and at different times, and here and there even interrupted by a local or temporary regression. This further development has been strongly urged forward, on the one hand, and has been guided along more definite directions on the other hand, owing to a new element which came into play with the appearance of fully-fledged man, viz. *society*.

Hundreds of thousands of years—of no greater significance in the history of the earth than one second in the life of man [1]—certainly elapsed before human society arose out of a band of tree-climbing monkeys. Yet it did finally appear. And what do we find once more as the characteristic difference between the band of monkeys and human society? *Labour*. The ape horde was satisfied to browse over the feeding area determined for it by geographical conditions or the degree of resistance of neighbouring hordes ; it undertook migrations and struggles to win new feeding grounds, but it was incapable of extracting from the area which supplied it with food more than the region offered in its natural state, except, perhaps, that the horde unconsciously fertilised the soil with its own excrements. As

---

[1] A leading authority in this respect, Sir W. Thomson, has calculated that *little more than a hundred million years* * could have elapsed since the time when the earth had cooled sufficiently for plants and animals to be able to live on it. [*Note by F. Engels.*]

* This time has been greatly extended by the discovery of radio-activity. The correct figure is probably about fifteen hundred million years.

soon as all possible feeding grounds were occupied, further increase of the monkey population could not occur ; the number of animals could at best remain stationary. But all animals waste a great deal of food, and, in addition, destroy in embryo the next generation of the food supply. Unlike the hunter, the wolf does not spare the doe which would provide it with young deer in the next year ; the goats in Greece, which graze down the young bushes before they can grow up, have eaten bare all the mountains of the country. This " predatory economy " of animals plays an important part in the gradual transformation of species by forcing them to adapt themselves to other than the usual food, thanks to which their blood acquires a different chemical composition and the whole physical constitution gradually alters,[1] while species that were once established die out. There is no doubt that this predatory economy has powerfully contributed to the gradual evolution of our ancestors into men. In a race of apes that far surpassed all others in intelligence and adaptability, this predatory economy could not help leading to a continual increase in the number of plants used for food and to the devouring of more and more edible parts of these plants. In short, it led to the food becoming more and more varied, hence also the substances entering the body, the chemical premises for the transition to man. But all that was not yet labour in the proper sense of the word. The labour process begins with the making of tools. And what are the most ancient tools that we find—the most ancient judging by the heir-looms of prehistoric man that have been discovered, and by the mode of life of the earliest historical peoples and of the most primitive of contemporary savages ? They are hunting and fishing implements, the former at

[1] It is very doubtful whether evolution occurs as a result of this process.

the same time serving as weapons. But hunting and
fishing presuppose the transition from an exclusively
vegetable diet to the concomitant use of meat, and this
is an important step in the transition to man. A *meat
diet* contains in an almost ready state the most essential
ingredients required by the organism for its metabolism.
It shortened the time required, not only for digestion,
but also for the other vegetative bodily processes
corresponding to those of plant life, and thus gained
further time, material, and energy for the active manifestation of animal life in the proper sense of the word.
And the further that man in the making became removed
from the plant kingdom, the higher he rose also over
animals. Just as becoming accustomed to a plant diet
side by side with meat has converted wild cats and
dogs into the servants of man, so also adaptation to a
flesh diet, side by side with a vegetable diet, has considerably contributed to giving bodily strength and
independence to man in the making. The most
essential effect, however, of a flesh diet was on the brain,
which now received a far richer flow of the materials
necessary for its nourishment and development, and
which therefore could become more rapidly and perfectly
developed from generation to generation.[1] With all
respect to the vegetarians, it has to be recognised that
man did not come into existence without a flesh diet,
and if the latter, among all peoples known to us, has led
to cannibalism at some time or another (the forefathers
of the Berliners, the Weletabians or Wilzians, used to eat
their parents as late as the tenth century), that is of no
consequence to us to-day.

A meat diet led to two new advances of decisive
importance : to the mastery of fire and the taming of

---

[1] Engels' belief in a meat diet is by no means shared universally
by students of biochemistry, although it must be remembered that most
so-called vegetarians partake of milk or its products.

animals. The first still further shortened the digestive process, as it provided the mouth with food already as it were semi-digested; the second made meat more copious by opening up a new, more regular source of supply in addition to hunting, and moreover provided, in milk and its products, a new article of food at least as valuable as meat in its composition. Thus, both these advances became directly new means of emancipation for man. It would lead us too far to dwell here in detail on their indirect effects notwithstanding the great importance they have had for the development of man and society.

Just as man learned to consume everything edible, he learned also to live in any climate. He spread over the whole of the habitable world, being the only animal that by its very nature had the power to do so. The other animals that have become accustomed to all climates — domestic animals and vermin — did not become so independently, but only in the wake of man. And the transition from the uniformly hot climate of the original home of man to colder regions, where the year is divided into summer and winter, created new requirements: shelter and clothing as protection against cold and damp, new spheres for labour and hence new forms of activity, which further and further separated man from the animal.

By the co-operation of hands, organs of speech, and brain, not only in each individual, but also in society, human beings became capable of executing more and more complicated operations, and of setting themselves, and achieving, higher and higher aims. With each generation,[1] labour itself became different, more perfect,

---

[1] This is probably an exaggeration. A study of stone-age technique suggests that periods of stagnation lasted for scores or hundreds of generations. Of course the time occupied by human evolution is much longer than Engels (or his scientific contemporaries) thought possible.

more diversified. Agriculture was added to hunting and cattle-breeding, then spinning, weaving, metalworking, pottery, and navigation. Along with trade and industry, there appeared finally art and science. From tribes there developed nations and states. Law and politics arose, and with them the fantastic reflection of human things in the human mind : religion. In the face of all these creations, which appeared in the first place to be products of the mind, and which seemed to dominate human society, the more modest productions of the working hand retreated into the background, the more so since the mind that plans the labour process already at a very early stage of development of society (*e.g.* already in the simply family), was able to have the labour that had been planned carried out by other hands than its own. All merit for the swift advance of civilisation was ascribed to the mind, to the development and activity of the brain. Men became accustomed to explain their actions from their thoughts, instead of from their needs—(which in any case are reflected and come to consciousness in the mind)—and so there arose in the course of time that idealistic outlook on the world which, especially since the decline of the ancient world, has dominated men's minds. It still rules them to such a degree that even the most materialistic natural scientists of the Darwinian school are still unable to form any clear idea of the origin of man, because under this ideological influence they do not recognise the part that has been played therein by labour.

Animals, as already indicated, change external nature by their activities just as man does, if not to the same extent, and these changes made by them in their environment, as we have seen, in turn react upon and change their originators. For in nature nothing takes place in isolation. Everything affects every other thing and *vice versa*, and it is usually because this many-sided motion

290 DIALECTICS OF NATURE

and interaction is forgotten that our natural scientists
are prevented from clearly seeing the simplest things.
We have seen how goats have prevented the regeneration
of forest in Greece ; on the island of St. Helena, goats and
pigs brought by the first arrivals have succeeded in
exterminating almost completely the old vegetation of
the island, and so have prepared the soil for the spreading
of plants brought by later sailors and colonists. But if
animals exert a lasting effect on their environment, it
happens unintentionally, and, as far as the animals
themselves are concerned, it is an accident. The
further men become removed from animals, however,
the more their effect on nature assumes the character
of a premeditated, planned action directed towards
definite ends known in advance. The animal destroys
the vegetation of a locality without realising what it is
doing. Man destroys it in order to sow field crops on
the soil thus released, or to plant trees or vines which
he knows will yield many times the amount sown. He
transfers useful plants and domestic animals from one
country to another and thus changes the flora and fauna
of whole continents. More than this. Under artificial
cultivation, both plants and animals are so changed by
the hand of man that they become unrecognisable.
The wild plants from which our grain varieties originated
are still being sought in vain.[1] The question of the wild
animal from which our dogs are descended, the dogs
themselves being so different from one another, or our
equally numerous breeds of horses, is still under dispute.

In any case, of course, we have no intention of
disputing the ability of animals to act in a planned and
premeditated fashion. On the contrary, a planned mode
of action exists in embryo wherever protoplasm, living
protein, exists and reacts, i.e. carries out definite, even
if extremely simple, movements as a result of definite

[1] They are now, in many cases, known with fair certainty.

external stimuli. Such reaction takes place even where there is as yet no cell at all, far less a nerve cell. The manner in which insectivorous plants capture their prey appears likewise in a certain respect as a planned action, although performed quite unconsciously. In animals the capacity for conscious, planned action develops side by side with the development of the nervous system and among mammals it attains quite a high level. While fox-hunting in England, one can daily observe how unerringly the fox knows how to make use of its excellent knowledge of the locality in order to escape from its pursuers, and how well it knows and turns to account all favourable features of the ground that cause the scent to be interrupted. Among our domestic animals, more highly developed thanks to association with man, every day one can note acts of cunning on exactly the same level as those of children. For, just as the developmental history of the human embryo in the mother's womb is only an abbreviated repetition of the history, extending over millions of years, of the bodily evolution of our animal ancestors, beginning from the worm, so the mental development of the human child is only a still more abbreviated repetition of the intellectual development of these same ancestors, at least of the later ones. But all the planned action of all animals has never resulted in impressing the stamp of their will upon nature. For that, man was required.

In short, the animal merely *uses* external nature, and brings about changes in it simply by his presence ; man by his changes makes it serve his ends, *masters* it. This is the final, essential distinction between man and other animals, and once again it is labour that brings about this distinction.

Let us not, however, flatter ourselves overmuch on account of our human conquest over nature. For each such conquest takes its revenge on us. Each of them, it is true, has in the first place the consequences on which we counted, but in the second and third places it has quite different, unforeseen effects which only too often cancel out the first. The people who, in Mesopotamia, Greece, Asia Minor, and elsewhere, destroyed the forests to obtain cultivable land, never dreamed that they were laying the basis for the present devastated condition of these countries, by removing along with the forests the collecting centres and reservoirs of moisture. When, on the southern slopes of the mountains, the Italians of the Alps used up the pine forests so carefully cherished on the northern slopes, they had no inkling that by doing so they were cutting at the roots of the dairy industry in their region; they had still less inkling that they were thereby depriving their mountain springs of water for the greater part of the year, with the effect that these would be able to pour still more furious flood torrents on the plains during the rainy seasons. Those who spread the potato in Europe were not aware that they were at the same time spreading the disease of scrofula.[1] Thus at every step we are reminded that we by no means rule over nature like a conqueror over a foreign people, like someone standing outside nature—but that we, with flesh, blood, and brain, belong to nature, and exist in its midst, and that all our mastery of it consists in the fact that we have the advantage over all other beings of being able to know and correctly apply its laws.

And, in fact, with every day that passes we are learning to understand these laws more correctly, and getting

[1] At the time when Engels wrote it was widely believed in medical circles that scrofula (tuberculosis of the neck glands) was due to eating potatoes. There is a causal connection in the sense that it is a disease of inadequately fed people, including those who live on a diet mainly of potatoes. But there is no real evidence that potatoes, as such, play any part in causing it.

to know both the more immediate and the more remote
consequences of our interference with the traditional
course of nature. In particular, after the mighty
advances of natural science in the present century,
we are more and more getting to know, and hence to
control, even the more remote natural consequences
at least of our more ordinary productive activities.
But the more this happens, the more will men not only
feel, but also know, their unity with nature, and thus
the more impossible will become the senseless and anti-
natural idea of a contradiction between mind and
matter, man and nature, soul and body, such as arose
in Europe after the decline of classic antiquity and which
obtained its highest elaboration in Christianity.

But if it has already required the labour of thousands
of years for us to learn to some extent to calculate the
more remote *natural* consequences of our actions aiming
at production, it has been still more difficult in regard
to the more remote *social* consequences of these actions.
We mentioned the potato and the resulting spread of
scrofula. But what is scrofula in comparison with the
effect on the living conditions of the masses of the
people in whole countries resulting from the workers
being reduced to a potato diet, or in comparison with
the famine which overtook Ireland in 1847 in consequence
of the potato disease, and which put under the earth a
million Irishmen, nourished solely or almost exclusively
on potatoes, and forced the emigration overseas of two
million more ? When the Arabs learned to distil
alcohol, it never entered their heads that by so doing
they were creating one of the chief weapons for the
annihilation of the original inhabitants of the still
undiscovered American continent. And when afterwards Columbus discovered America, he did not know
that by doing so he was giving new life to slavery,
which in Europe had long ago been done away with,

and laying the basis for the Negro slave traffic. The men who in the seventeenth and eighteenth centuries laboured to create the steam engine had no idea that they were preparing the instrument which more than any other was to revolutionise social conditions throughout the world. Especially in Europe, by concentrating wealth in the hands of a minority, the huge majority being rendered propertyless, this instrument was destined at first to give social and political domination to the bourgeoisie, and then, however, to give rise to a class struggle between bourgeoisie and proletariat, which can end only in the otherthrow of the bourgeoisie and the abolition of all class contradictions. But even in this sphere, by long and often cruel experience and by collecting and analysing the historical material, we are gradually learning to get a clear view of the indirect, more remote, social effects of our productive activity, and so the possibility is afforded us of mastering and controlling these effects as well.

To carry out this control requires something more than mere knowledge. It requires a complete revolution in our hitherto existing mode of production, and with it of our whole contemporary social order.

All hitherto existing modes of production have aimed merely at achieving the most immediately and directly useful effect of labour. The further consequences, which only appear later on and become effective through gradual repetition and accumulation, were totally neglected. Primitive communal ownership of land corresponded, on the one hand, to a level of development of human beings in which their horizon was restricted in general to what lay immediately at hand, and presupposed, on the other hand, a certain surplus of available land, allowing a certain latitude for correcting any possible bad results of this primitive forest type of economy. When this surplus land was exhausted, communal ownership also declined. All higher forms of production, however, proceeded in their development to the division of the population into different classes and thereby to the contradiction of ruling and oppressed classes. But thanks to this, the interest of the ruling class became the driving factor of production, in so far as the latter was not restricted to the barest means of subsistence of the oppressed people. This has been carried through most completely in the capitalist mode of production prevailing to-day in Western Europe. The individual capitalists, who dominate production and exchange, are able to concern themselves only with the most immediate useful effect of their actions. Indeed, even this useful effect—in as much as it is a question of the usefulness of the commodity that is produced or exchanged—retreats right into the background, and the sole incentive becomes the profit to be gained on selling.

The social science of the bourgeoisie, classical political economy, is predominantly occupied only with the directly intended social effects of human actions connected with production and exchange. This fully corresponds to the social organisation of which it is the theoretical expression. When individual capitalists are engaged in production and exchange for the sake of the immediate profit, only the nearest, most immediate results can be taken into account in the first place. When an individual manufacturer or merchant sells a manufactured or purchased commodity with only the usual small profit, he is satisfied, and he is not concerned as to what becomes of the commodity afterwards or who are its purchasers. The same thing applies to the natural effects of the same actions. What did the Spanish planters in Cuba, who burned down forests on the slopes of the mountains and obtained from the ashes sufficient fertiliser for *one* generation of very highly profitable coffee trees, care that the tropical rainfall afterwards washed away the now unprotected upper stratum of the soil, leaving behind only bare rock ?   In relation to nature, as to society, the present mode of production is predominantly concerned only about the first, tangible success ;   and then surprise is expressed that the more remote effects of actions directed to this end turn out to be of quite a different, mainly even of quite an opposite, character ; that the harmony of demand and supply becomes transformed into their polar opposites, as shown by the course of each ten years' industrial cycle, and of which even Germany has experienced a little preliminary in the " crash " ; that private ownership based on individual labour necessarily develops into the propertylessness of the workers, while all wealth becomes more and more concentrated in the hands of non-workers ; that . . .[1]

[1] The manuscript here breaks off abruptly.

X
NATURAL SCIENCE AND THE SPIRIT WORLD [1]

THE dialectics that has found its way into popular consciousness finds expression in the old saying that extremes meet. In accordance with this we should hardly err in looking for the most extreme degree of fantasy, credulity, and superstition, not in that trend of natural science which, like the German philosophy of nature, tries to force the objective world into the framework of its subjective thought, but rather in the opposite trend, which, relying on mere experience, treats thought with sovereign disdain and really has gone to the furthest extreme in emptiness of thought. This school prevails in England. Its father, the much lauded Francis Bacon, already advanced the demand that his new empirical-inductive method should be pursued to attain by its means, above all, longer life, rejuvenation—to a certain extent, alteration of stature and features, transformation of one body into another, the production of new species, power over the air and the production of storms. He complains that such investigations have been abandoned, and in his natural history he actually gives recipes for making gold and performing various miracles. Similarly Isaac Newton in his old age greatly busied himself with expounding the revelation of St. John. So it is not to be wondered at if in recent years English empiricism in the person of some of its representatives—and not the worst of them—

[1] From a manuscript of Engels probably written in 1878, and first published in the " *Illustrierter Neue Welt-Kalender für das Jahr 1898.*"

should seem to have fallen a hopeless victim to the spirit-rapping and spirit-seeing imported from America.

The first natural scientist belonging here is the very eminent zoologist and botanist, Alfred Russell Wallace, the man who simultaneously with Darwin put forward the theory of the evolution of species by natural selection. In his little work, *On Miracles and Modern Spiritualism*, London, Burns, 1875, he relates that his first experiences in this branch of natural knowledge date from 1844, when he attended the lectures of Mr. Spencer Hall on mesmerism and as a result carried out similar experiments on his pupils. " I was extremely interested in the subject and pursued it with ardour." He not only produced magnetic sleep together with the phenomena of articular rigidity and local loss of sensation, he also confirmed the correctness of Gall's map of the skull, because on touching any one of Gall's organs the corresponding activity was aroused in the magnetised patient and exhibited by appropriate and lively gestures. Further, he established that his patient, merely by being touched, partook of all the sensations of the operator ; he made him drunk with a glass of water as soon as he told him that it was brandy. He could make one of the young men so stupid, even in the waking condition, that he no longer knew his own name, a feat, however, that other schoolmasters are capable of accomplishing without any mesmerism. And so on.

Now it happens that I also saw this Mr. Spencer Hall in the winter of 1843–4 in Manchester. He was a very mediocre charlatan, who travelled the country under the patronage of some parsons and undertook magnetico-phrenological performances with a young girl in order to prove thereby the existence of God, the immortality of the soul, and the incorrectness of the materialism that was being preached at that time by the Owenites in all big towns. The lady was sent into a magnetico-sleep and then, as soon as the operator touched any part of the skull corresponding to one of Gall's organs, she gave a bountiful display of theatrical, demonstrative gestures and poses representing the activity of the organ concerned ; for instance, for the organ of philoprogenitiveness she fondled and kissed an imaginary baby, etc. Moreover, the good Mr. Hall had enriched Gall's geography of the skull with a new island of Barataria : right at the top of the skull he had discovered an organ of veneration, on touching which his hypnotic miss sank on to her knees, folded her hands in prayer, and depicted to the astonished, philistine audience an angel wrapt in veneration. That was the climax and conclusion of the exhibition. The existence of God had been proved.

The effect on me and one of my acquaintances was exactly the same as on Mr. Wallace ; the phenomena interested us and we tried to find out how far we could reproduce them. A wideawake young boy of 12 years old offered himself as subject. Gently gazing into his eyes, or stroking, sent him without difficulty into the hypnotic condition. But since we were rather less credulous than Mr. Wallace and set to work with rather less fervour, we arrived at quite different results. Apart from muscular rigidity and loss of sensation, which were easy to produce, we found also a state of complete passivity of the will bound up with a peculiar hypersensitivity of sensation. The patient, when aroused from his lethargy by any external stimulus, exhibited very much greater liveliness than in the waking condition. There was no trace of any mysterious relation to the operator ; anyone else could just as easily set the sleeper into activity. To set Gall's cranial organs into action was the least that we achieved ; we went much further, we could not only exchange them for one another, or make their seat anywhere in the whole body, but we also fabricated any amount of other organs, organs of singing, whistling, piping, dancing,

boxing, sewing, cobbling, tobacco-smoking, etc., and
we could make their seat wherever we wanted. Wallace
made his patients drunk on water, but we discovered
in the great toe an organ of drunkenness which only had
to be touched in order to cause the finest drunken
comedy to be enacted. But it must be well understood,
no organ showed a trace of action until the patient was
given to understand what was expected of him ; the
boy soon perfected himself by practice to such an extent
that the merest indication sufficed. The organs produced in this way then retained their validity for later
occasions of putting to sleep, as long as they were not
altered in the same way. The patient had even a double
memory, one for the waking state and a second quite
separate one for the hypnotic condition. As regards the
passivity of the will and its absolute subjection to the
will of a third person, this loses all its miraculous appearance when we bear in mind that the whole condition
began with the subjection of the will of the patient to
that of the operator, and cannot be restored without it.
The most powerful magician of a magnetiser in the world
will come to the end of his resources as soon as his patient
laughs him in the face.

While we with our frivolous scepticism thus found
that the basis of magnetico-phrenological charlatanry
lay in a series of phenomena which for the most part
differ only in degree from those of the waking state and
require no mystical interpretation, Mr. Wallace's
" ardour " led him into a series of self-deceptions, in
virtue of which he confirmed Gall's map of the skull in
all its details and noted a mysterious relation between
operator and patient.[1] Everywhere in Mr. Wallace's

[1] As already said, the patients perfect themselves by practice.
It is therefore quite possible that, when the subjection of the will has
become habitual, the relation of the participants becomes more intimate,
individual phenomena are intensified and are reflected weakly even
in the waking state. [Note by F. Engels.]

account, the sincerity of which reaches the degree of naiveté, it becomes apparent that he was much less concerned in investigating the factual background of charlatanry than in reproducing all the phenomena at all costs. Only this frame of mind is needed for the man who was originally a scientist to be quickly converted into an " adept " by means of simple and facile self-deception. Mr. Wallace ended up with faith in magnetico-phrenological miracles and so already stood with one foot in the world of spirits.

He drew the other foot after him in 1865. On returning from his twelve years of travel in the tropical zone, experiments in table-turning introduced him to the society of various " mediums." How rapid his progress was, and how complete his mastery of the subject, is testified to by the above-mentioned booklet. He expects us to take for good coin not only all the alleged miracles of Home, the brothers Davenport, and other " mediums " who all more or less exhibit themselves for money and who have for the most part been frequently exposed as impostors, but also a whole series of allegedly authentic spirit histories from early times. The Pythonesses of the Greek oracle, the witches of the Middle Ages, were all " mediums," and Iamblichus [1] in his *De divinatione* already described quite accurately " the most astonishing phenomena of modern spiritualism."

Just one example to show how lightly Mr. Wallace deals with the scientific corroboration and authentication of these miracles. It is certainly a strong assumption that we should believe that the aforesaid spirits should allow themselves to be photographed, and we have surely the right to demand that such spirit photographs should be authenticated in the most indubitable manner before we accept them as genuine. Now Mr. Wallace

[1] See Appendix II, p. 368.

302 DIALECTICS OF NATURE

recounts on p. 187 that in March, 1872, a leading medium, Mrs. Guppy, *née* Nicholls, had herself photographed together with her husband and small boy at Mr. Hudson's in Notting Hill, and on two different photographs a tall female figure, finely draped in white gauze robes, with somewhat Eastern features, was to be seen behind her in a pose as if giving a benediction. " Here, then, one of two things are absolutely certain.[1] Either there was a living intelligent, but invisible being present, or Mr. and Mrs. Guppy, the photographer, and some fourth person planned a wicked imposture and have maintained it ever since. Knowing Mr. and Mrs. Guppy so well as I do, I feel an *absolute conviction* that they are as incapable of an imposture of this kind as any earnest inquirer after truth in the department of natural science." [2]

Consequently, either deception or spirit photography. Quite so. And, if deception, either the spirit was already on the photographic plates, or four persons must have been concerned, or three if we leave out as weak-minded or duped old Mr. Guppy who died in January, 1875, at the age of 84 (it only needed that he should be sent behind the Spanish screen of the background). That a photographer could obtain a " model " for the spirit without difficulty does not need to be argued. But the photographer Hudson, shortly afterwards, was publicly prosecuted for habitual falsification of spirit photographs, so Mr. Wallace remarks in mitigation : " One thing is clear, if an imposture has occurred, it was at once detected by spiritualists themselves." Hence there is not much reliance to be placed on the photographer. Remains Mrs. Guppy, and for her there is

---

[1] The spirit world is superior to grammar. A joker once caused the spirit of the grammarian Lindley Murray to testify. To the question whether he was there, he answered : " I are." (American for I am.) The medium was from America. [*Note by F. Engels.*]

[2] See Appendix II, p. 369.

only the " absolute conviction " of our friend Wallace and nothing more. Nothing more ? Not at all. The absolute trustworthiness of Mrs. Guppy is evidenced by her assertion that one evening, early in June, 1871, she was carried through the air in a state of unconsciousness from her house in Highbury Hill Park to 69, Lamb's Conduit Street—three English miles as the crow flies— and deposited in the said house of No. 69 on the table in the midst of a spiritualistic séance. The doors of the room were closed, and although Mrs. Guppy was one of the stoutest women in London, which is certainly saying a good deal, nevertheless her sudden incursion did not leave behind the slightest hole either in the doors or in the ceiling. (Reported in the London *Echo*, June 8, 1871.) And if anyone still does not believe in the genuineness of spirit photography, there's no helping him.

The second eminent adept among English natural scientists is Mr. William Crookes, the discoverer of the chemical element thallium and of the radiometer (in Germany also called " *Lichtmühle* " [light-mill]). Mr. Crookes began to investigate spiritualistic manifestations about 1871, and employed for this purpose a number of physical and mechanical appliances, spring balances, electric batteries, etc. Whether he brought to his task the main apparatus required, a sceptically critical mind, or whether he remained to the end in a fit state for working, we shall see. At any rate, within a not very long period, Mr. Crookes was just as completely captivated as Mr. Wallace. " For some years," he relates, " a young lady, Miss Florence Cook, has exhibited remarkable mediumship, which latterly culminated in the production of an entire female form purporting to be of spiritual origin, and which appeared barefooted and in white flowing robes while she lay entranced in dark clothing and securely bound in a

304 DIALECTICS OF NATURE

cabinet or adjoining room." This spirit, which called itself Katie, and which looked remarkably like Miss Cook, was one evening suddenly seized round the waist by Mr. Volckmann—the present husband of Mrs. Guppy—and held fast in order to see whether it was not indeed Miss Cook in another edition. The spirit proved to be a quite sturdy damsel, it defended itself vigorously, the onlookers intervened, the gas was turned out, and when, after some scuffling, peace was re-established and the room re-lit, the spirit had vanished and Miss Cook lay bound and unconscious in her corner. Nevertheless, Mr. Volckmann is said to maintain up to the present day that he had seized hold of Miss Cook and nobody else. In order to establish this scientifically, Mr. Varley, a well-known electrician, on the occasion of a new experiment, arranged for the current from a battery to flow through the medium, Miss Cook, in such a way that she could not play the part of the spirit without interrupting the current. Nevertheless, the spirit made its appearance. It was, therefore, indeed a being different from Miss Cook. To establish this further was the task of Mr. Crookes. His first step was to win the *confidence* of the spiritualistic lady. This confidence, so he says himself in the *Spiritualist*, June 5, 1874, " increased gradually to such an extent that she refused to give a séance unless *I made the arrangements*. She said that she always wanted *me* to be near her and in the neighbourhood of the cabinet; I found that— when this confidence had been established and she was sure that *I would not break any promise made to her*—the phenomena increased considerably in strength and there was freely forthcoming evidence that would have been unobtainable in any other way. She frequently *consulted me* in regard to the persons present at the seances and the places to be given them, for she had recently become very nervous as a result of certain ill-advised suggestions that, besides other more scientific methods of investigation, *force* also should be applied."

The spirit lady rewarded this confidence, which was as kind as it was scientific, in the highest measure. She even made her appearance—which can no longer surprise us—in Mr. Crookes' house, played with his children and told them " anecdotes from her adventures in India," treated Mr. Crookes to an account of " some of the bitter experiences of her past life," allowed him to take her by the arm so that he could convince himself of her evident materiality, allowed him to take her pulse and count the number of her respirations per minute, and finally allowed herself to be photographed next to Mr. Crookes. " This figure," says Mr. Wallace, " after she had been seen, touched, photographed, and conversed with, *vanished absolutely* out of a small room from which there was no other exit than an adjoining room filled with spectators "—which was not such a great feat, provided that the spectators were polite enough to show as much faith in Mr. Crookes, in whose house this happened, as Mr. Crookes did in the spirit.

Unfortunately these "fully authenticated phenomena " are not immediately credible even for spiritualists. We saw above how the very spiritualistic Mr. Volckmann permitted himself to make a very material grab. And now a clergyman, a member of the committee of the " British National Association of Spiritualists," has also been present at a séance with Miss Cook, and he established the fact without difficulty that the room through the door of which the spirit came and disappeared communicated with the outer world by a *second door*. The behaviour of Mr. Crookes, who was also present, gave " the final death blow to my belief that there might be something in the manifestations." (*Mystic London,* by the Rev. C. Maurice Davies, London,

Tinsley Brothers).[1]   And, over and above that, it came
to light in America how " Katies " were " materialised."
A married couple named Holmes held séances in Philadelphia in which likewise a " Katie " appeared and
received bountiful presents from the believers.   However, one sceptic refused to rest until he got on the track
of the said Katie, who, anyway, had already gone on
strike once because of lack of pay ; he discovered her in a
boarding-house as a young lady of unquestionable
flesh and bone, and in possession of all the presents that
had been given to the spirit.

Meanwhile the Continent also had its scientific spirit-
seers. A scientific association at St. Petersburg—I
do not know exactly whether the University or even the
Academy itself—charged the Councillor of State,
Aksakov, and the chemist, Butlerov, to examine the
basis of the spiritualistic phenomena, but it does not
seem that very much came of this.   On the other hand
—if the noisy announcements of the spiritualists are to
be believed—Germany has now also put forward its
man in the person of Professor Zöllner in Leipzig.

For years, as is well known, Herr Zöllner has been
hard at work on the " fourth dimension " of space, and
has discovered that many things that are impossible in a
space of three dimensions, are a simple matter of course
in a space of four dimensions.   Thus, in the latter kind
of space, a closed metal sphere can be turned inside out
like a glove, without making a hole in it ; similarly a
knot can be tied in an endless string or one which has
both ends fastened, and two separate closed rings can be
interlinked without opening either of them, and many
more such feats.   According to the recent triumphant
reports from the spirit world, it is said now that Professor Zöllner has addressed himself to one or more
mediums in order with their aid to determine more

[1] See Appendix II, p. 370.

details of the locality of the fourth dimension. The success is said to have been surprising. After the session the arm of the chair, on which he rested his arm while his hand never left the table, was found to have become interlocked with his arm, a string that had both ends sealed to the table was found tied into four knots, and so on. In short, all the miracles of the fourth dimension are said to have been performed by the spirits with the utmost ease. It must be borne in mind : *relata refero*, I do not vouch for the correctness of the spirit bulletin, and if it should contain any inaccuracy, Herr Zöllner ought to be thankful that I am giving him the opportunity to make a correction. If, however, it reproduces the experiences of Herr Zöllner without falsification, then it obviously signifies a new era both in the science of spiritualism and that of mathematics. The spirits prove the existence of the fourth dimension, just as the fourth dimension vouches for the existence of spirits. And this once established, an entirely new, immeasurable field is opened to science. All previous mathematics and natural science will be only a preparatory school for the mathematics of the fourth and still higher dimensions, and for the mechanics, physics, chemistry, and physiology of the spirits dwelling in these higher dimensions. Has not Mr. Crookes scientifically determined how much weight is lost by tables and other articles of furniture on their passage into the fourth dimension—as we may now well be permitted to call it—and does not Mr. Wallace declare it proven that fire there does no harm to the human body ? And now we have even the physiology of the spirit bodies ! They breathe, they have a pulse, therefore lungs, heart, and a circulatory apparatus, and in consequence are at least as admirably equipped as our own in regard to the other bodily organs. For breathing requires carbohydrates which undergo combustion in the lungs, and

these carbohydrates can only be supplied from without ;
hence, stomach, intestines, and their accessories—and
if we have once established so much, the rest follows
without difficulty. The existence of such organs,
however, implies the possibility of their falling a prey to
disease, hence it may still come to pass that Herr
Virchow will have to compile a cellular pathology of
the spirit world. And since most of these spirits are
very handsome young ladies, who are not to be distinguished in any respect whatsoever from terrestrial
damsels, other than by their supra-mundane beauty,
it could not be very long before they come into contact
with " men who feel the passion of love " ; and since,
as established by Mr. Crookes from the beat of the pulse,
" the female heart is not absent," natural selection also
has opened before it the prospect of a fourth dimension,
one in which it has no longer any need to fear of being
confused with wicked social-democracy.

Enough. Here it becomes palpably evident which is
the most certain path from natural science to mysticism.
It is not the extravagant theorising of the philosophy of
nature, but the shallowest empiricism that spurns all
theory and distrusts all thought. It is not *a priori*
necessity that proves the existence of spirits, but the
empirical observations of Messrs. Wallace, Crookes,
and Co. If we trust the spectrum-analysis observations
of Crookes, which led to the discovery of the metal
thallium, or the rich zoological discoveries of Wallace in
the Malay Archipelago, we are asked to place the same
trust in the spiritualistic experiences and discoveries of
these two scientists. And if we express the opinion
that, after all, there is a little difference between the
two, namely, that we can verify the one but not the
other, then the spirit-seers retort that this is not the

case, and that they are ready to give us the opportunity of verifying also the spirit phenomena.

Indeed, dialectics cannot be despised with impunity. However great one's contempt for all theoretical thought, nevertheless one cannot bring two natural facts into relation with one another, or understand the connection existing between them, without theoretical thought. The only question is whether one's thinking is correct or not, and contempt of theory is evidently the most certain way to think naturalistically, and therefore incorrectly. But, according to an old and well-known dialectical law, incorrect thinking, carried to its logical conclusion, inevitably arrives at the opposite of its point of departure. Hence, the empirical contempt of dialectics on the part of some of the most sober empiricists is punished by their being led into the most barren of all superstitions, into modern spiritualism.

It is the same with mathematics. The ordinary, metaphysical mathematicians boast with enormous pride of the absolute irrefutability of the results of their science. But these results include also imaginary magnitudes, which thereby acquire a certain reality. When one has once become accustomed to ascribe some kind of reality outside of our minds to $\sqrt{-1}$, or to the fourth dimension, then it is not a matter of much importance if one goes a step further and also accepts the spirit world of the mediums. It is as Ketteler said about Döllinger [1] : " The man has defended so much nonsense in his life, he really could have accepted infallibility into the bargain ! "

In fact, mere empiricism is incapable of refuting the spiritualists. In the first place, the " higher " phenomena always show themselves only when the " investigator " concerned is already so far in the toils

---

[1] A catholic scholar who did not accept the dogma of papal infallibility.

that he now only sees what he is meant to see or wants to see—as Crookes himself describes with such inimitable naïveté. In the second place, however, the spiritualist cares nothing that hundreds of alleged facts are exposed as imposture and dozens of alleged mediums as ordinary tricksters. As long as *every* single alleged miracle has not been explained away, they have still room enough to carry on, as indeed Wallace says clearly enough in connection with the falsified spirit photographs. The existence of falsifications proves the genuineness of the genuine ones.

And so empiricism finds itself compelled to refute the importunate spirit-seers not by means of empirical experiments, but by theoretical considerations, and to say, with Huxley[1] : " The only good that I can see in the demonstration of the truth of ' spiritualism ' is to furnish an additional argument against suicide. Better live a crossing-sweeper than die and be made to talk twaddle by a ' medium '. hired at a guinea a séance ! "

[1] See Appendix II, p. 370.

APPENDICES

APPENDIX I

NOTES TO "ANTI-DÜHRING"

The following notes were written by Engels towards the end of 1877 and beginning of 1878, after the publication in separate form of the first section (Philosophy) of " Anti-Dühring," to the pages of which he refers at the beginning of each note. In view of their great intrinsic importance and their close connection with the subjects dealt with in Dialectics of Nature, they are included here as an appendix.

(a) ON THE PROTOTYPES OF MATHEMATICAL "INFINITY" IN THE REAL WORLD.

> *Re pp. 17–18: Concordance of thought and being— Mathematical infinity.*

The fact that our subjective thought and the objective world are subject to the same laws, and that consequently too in the final analysis they cannot be in contradiction to one another in their results, but must coincide, governs absolutely our whole theoretical thought. It is the unconscious and unconditional premise for theoretical thought. Eighteenth century materialism, owing to its essentially metaphysical character, investigated this premise only as regards content. It restricted itself to the proof that the content of all thought and knowledge must derive from sensuous experience, and revived the principle : *nihil est in intellectu, quod non fuerit in sensu.* It was modern idealistic, but at the same time dialectical, philosophy, and especially Hegel, which for the first time investigated it also as regards *form.* In spite of all the innumerable arbitrary constructions and fantasies that we encounter here, in spite of the idealist, topsy-turvy, form of its result—the unity of thought and being—it is undeniable that this philosophy proved the analogy of the processes of thought to those of nature and history and *vice versa,* and the validity of similar laws for all these processes, in numerous cases and in the most diverse fields. On the other hand, modern natural science has extended the principle of the origin of all thought content from experience in a way that breaks down its old metaphysical limitation and formulation. By recognising the inheritance of acquired characters, it extends the subject of experience from the individual to the genus ; the single individual that must have experienced is no longer necessary, its individual experience can be replaced to a certain extent by the results of the experiences of a number of its ancestors. If, for instance, among us the mathematical axioms seem self-evident to every eight-year-old child, and in no need of proof from experience, this is solely the result of " accumulated inheritance." It would be difficult to teach them by a proof to a bushman or Australian negro.

In the present work dialectics is conceived as the science of the most general laws of *all* motion. Therein is included that their laws must be equally valid for motion in nature and human history and for the motion of thought. Such a law can be recognised in two of these three spheres, indeed even in all three, without the metaphysical philistine being clearly aware that it is one and the same law that he has come to know.

Let us take an example. Of all theoretical advances there is surely none that ranks so high as a triumph of the human mind as the discovery of the infinitesimal calculus in the last half of the seventeenth century. If anywhere, it is here that we have a pure and exclusive feat of human intelligence. The mystery which even to-day surrounds the magnitudes employed in the infinitesimal calculus, the differentials and infinites of various degree, is the best proof that it is still imagined that what are dealt with here are pure " free creations and imaginings " of the human mind, to which there is nothing corresponding in the objective world. Yet the contrary is the case. Nature offers prototypes for all these imaginary magnitudes.

Our geometry has, as its starting point, space relations, and our arithmetic and algebra numerical magnitudes, which correspond to our terrestrial conditions, which therefore correspond to the magnitude of bodies that mechanics terms masses—masses such as occur on earth and are moved by men. In comparison to these masses, the mass of the earth seems infinitely large and indeed terrestrial mechanics treats it as infinitely large. The radius of the earth $= \infty$, this is the basic principle of all mechanics in the law of falling. But not merely the earth but the whole solar system and the distances occurring in the latter in their turn appear infinitely small as soon as we have to deal with the distances reckoned in light years in the stellar system visible to us through the telescope. We have here, therefore, already an infinity, not only of the first but of the second degree, and we can leave it to the imagination of our readers to construct further infinities of a higher degree in infinite space, if they feel inclined to do so.

According to the view prevailing in physics and chemistry to-day, however, the terrestrial masses, the bodies with which mechanics operates, consists of molecules, of smallest particles which cannot be further divided without abolishing the physical and chemical identity of the body concerned. According to W. Thomson's calculations, the diameter of the smallest of these molecules cannot be smaller than a fifty-millionth of a millimetre. But even if we assume that the largest molecule itself attains a diameter of a twenty-five-millionth of a millimetre, it still remains an infinitesimally small magnitude compared with the smallest mass dealt with by mechanics, physics, or even chemistry. Nevertheless, it is endowed with all the properties peculiar to the mass in question, it can represent the mass physically and chemically, and does actually represent it in all chemical equations. In short, it has the same properties in relation to the corresponding mass as the mathematical differential has in relation to its variable. The only difference is that what seems mysterious and inexplicable to us in the case of the differential, here seems a matter of course and as it were obvious.

Nature operates with these differentials, the molecules, in exactly the same way and according to the same laws as mathematics does with its abstract differentials. Thus, for instance, the differential of $x^3 = 3x^2 dx$, where $3x dx^2$ and dx^3 are neglected. If we put this in geometrical form, we have a cube with sides of length x, the length being increased by the infinitely small amount dx. Let us suppose that this cube consists of a sublimated element, say sulphur ; and that three of the surfaces around one corner are protected, the other three being free. Let us now expose this sulphur cube to an atmosphere of sulphur vapour and lower the temperature sufficiently ; sulphur will be deposited on the three free sides of the cube. We remain quite within the ordinary mode of procedure of physics and chemistry in supposing, in order to picture the process in its pure form, that in the first place a layer of thickness of a single molecule is deposited on each of these three sides. The length x of the sides of the cubes will have increased by the diameter of a molecule dx. The content of the cube x^3 has increased by the difference between x^3 and $x^3 + 3x^2 dx + 3x dx^2 + dx^3$, where dx^3, a *single* molecule and $3x dx^2$, three rows of length $x + dx$, consisting merely of lineally arranged molecules, can be neglected with the same justification as in mathematics. The result is the same, the increase in mass of the cube is $3x^2 dx$.

Strictly speaking dx^3 and $3x dx^2$ do not occur in the case of the sulphur molecule, because two or three molecules cannot occupy the same space, and the cube's increase of bulk is therefore exactly $3x^2 dx + 3x dx + dx$. This is explained by the fact that in mathematics dx is a linear magnitude, while it is well known that such lines, without thickness or breadth, do not occur independently in nature, hence also the mathematical abstractions have unrestricted validity only in pure mathematics. And since the latter neglects $3x dx^2 + dx^3$, it makes no difference.

Similarly in evaporation. When the uppermost molecular layer in a glass of water evaporates, the height of the water layer, x, is decreased by dx, and the continual flight of one molecular layer after another is actually a continued differentiation. And when the warm vapour is once more condensed to water in a vessel by pressure and cooling, and one molecular layer is deposited on another (it is permissible to leave out of account secondary circumstances that make the process an impure one) until the vessel is full, then literally an integration has been performed which differs from the mathematical one only in that the one is consciously carried out by the human brain, while the other is unconsciously carried out by nature. But it is not only in a transition from the liquid to the gaseous state and *vice versa* that processes occur which are completely analogous to those of the infinitesimal calculus.

When mass motion, as such, is abolished—by impact— and becomes transformed into heat, molecular motion, what is it that happens but that the mass motion is differentiated ? And when the movements of the molecules of steam in the cylinder of the steam engine become added together so that they lift the piston by a definite amount, so that they become transformed into mass motion, have they not been integrated ? Chemistry dissociates the molecules into atoms, magnitudes of more minute mass and spatial extension, but magnitudes of the same order, so that the two stand in definite, finite relations to one another. Hence, all the chemical equations which express the molecular composition of bodies are in their form differential equations. But in reality they are already integrated in the atomic weights which figure in them. For chemistry calculates with differentials, the mutual proportions of their magnitudes being known.

Atoms, however, are in no wise regarded as simple, or in general as the smallest known particles of matter. Apart from chemistry itself, which is more and more inclining to the view that atoms are compound, the majority of physicists assert that the luminiferous ether, which transmits light and heat radiations, likewise consists of discrete particles, which, however, are so small that they have the same relation to chemical atoms and physical molecules as these have to mechanical masses, that is to say as d^2x to dx. Here, therefore, in the now usual notion of the constitution of matter, we have likewise a differential of the second degree, and there is no reason at all why anyone, to whom it would give satisfaction, should not imagine that analogies of d^3x, d^4x, etc., also occur in nature.

Hence, whatever view one may hold of the constitution of matter, this much is certain, that it is divided up into a series of big, well-defined groups of a relatively massive character in such a way that the members of each separate group stand to one another in definite finite mass ratios, in contrast to which those of the next group stand to them in the ratio of the infinitely large or infinitely small in the mathematical sense. The visible system of stars, the solar system, terrestrial masses, molecules and atoms, and finally ether particles, each of them form such a group. It does not alter the case that intermediate links can be found between the separate groups. Thus, between the masses of the solar system and terrestrial masses come the asteroids (some of which have a diameter no greater than, for example, that of the Reuss principality, younger branch), meteors, etc. Thus, in the organic world the cell stands between terrestrial masses and molecules. These intermediate links prove only that there is no leap in nature, *precisely because* nature is composed entirely of leaps.

In so far as mathematics calculates with real magnitudes, it also employs this mode of outlook without hesitation. For terrestrial mechanics the mass of the earth is regarded as infinitely large, just as for astronomy terrestrial masses and the corresponding masses of meteors are regarded as infinitely small, and just as the distances and masses of the planets of the solar system are reduced to nothing as soon as astronomy investigates the constitution of our system of stars extending beyond the nearest fixed stars. As soon, however, as the mathematicians withdraw into their impregnable fortress of abstraction, so-called pure mathematics, all these analogies are forgotten, infinity becomes something totally mysterious, and the manner in which operations are carried out with it in analysis appears as something absolutely incomprehensible, contradicting all experience and all reason. The stupidities and absurdities by which mathematicians have rather excused than explained their mode of procedure, which remarkably enough always leads to correct results, exceed the most pronounced apparent and real fantasies, *e.g.* of the Hegelian philosophy of nature, about which mathematicians and natural scientists can never adequately express their horror. What they charge Hegel with doing, viz. pushing abstractions to the extreme limit, they do themselves on a far greater scale. They forget that the whole of so-called pure mathematics is concerned with abstractions, that *all* their magnitudes, taken in a strict sense, are imaginary, and that all abstractions when pushed to extremes are transformed into nonsense or into their opposite. Mathematical infinity is taken from reality although unconsciously, and consequently also can only be explained from reality and not from itself, from mathematical abstraction. And, as we have seen, if we investigate reality in this regard we come also upon the real relations from which the mathematical relation of infinity is taken, and even the natural analogies of the way in which this relation operates. And thereby the matter is explained. (Hæckel's bad reproduction of the identity of thinking and being.) But also *the contradiction between continuous and discrete matter*, see Hegel.

(b) On the " Mechanical " Conception of Nature.

Note 2. Re page 46 : *The various forms of motion and the sciences dealing with them.*

Since the above article appeared (*Vorwärts*, Feb. 9, 1877), Kekulé (*Die wissenschaftlichen Ziele und Leistungen der Chemie* [*The Scientific Aims and Achievements of Chemistry*] has defined mechanics, physics, and chemistry in a very similar way :

" If this idea of the nature of matter is made the basis, one could define chemistry as the science of atoms and physics as the science of molecules, and then it would be natural to separate that part of modern physics which deals with masses as a special science, reserving for it the name of mechanics. Thus mechanics appears as the basic science of physics and chemistry, in so far as in certain aspects and especially in certain calculations both of these have to treat their molecules or atoms as masses."

It will be seen that this formulation differs from that in the text and in the previous note only by being rather less definite. But when an English journal (*Nature*) [1] translated the above statement of Kekulé to the effect that mechanics is the statics and dynamics of masses, physics the statics and dynamics of molecules, and chemistry the statics and dynamics of atoms, then it seems to me that this unconditional reduction of even chemical processes to some-

---

[1] See quotation in Appendix II, p. 329.

thing merely mechanical unduly restricts the field, at least of chemistry. And yet it is so much the fashion that, for instance, Hæckel continually uses " mechanical " and " monistic " as having the same meaning, and in his opinion " modern physiology . . . in its field allows only of the operation of physico-chemical—or in the wider sense, mechanical—forces." (*Perigenesis*.[1])

If I term physics the mechanics of molecules, chemistry the physics of atoms, and furthermore biology the chemistry of proteins, I wish thereby to express the transition of each of these sciences into the other, hence both the connection, the continuity, and the distinction, the discrete separation. To go further and to define chemistry as likewise a kind of mechanics seems to me inadmissible. Mechanics— in the broader or narrower sense—knows only quantities, it calculates with velocities and masses, and at most with volumes. When the quality of bodies comes across its path, as in hydrostatics and aerostatics, it cannot achieve anything without going into molecular states and molecular motion, it is itself only a mere auxiliary science, the prerequisite for physics. In physics, however, and still more in chemistry, not only does continual qualitative change take place in consequence of quantitative change, the transformation of quantity into quality, but there are also many qualitative changes to be taken into account whose dependence on quantitative change is by no means proven. That the present tendency of science goes in this direction can be readily granted, but does not prove that this direction is the exclusively correct one, that the pursuit of this tendency will exhaust the whole of physics and chemistry. All motion includes mechanical motion, change of place of the largest or smallest portions of matter, and the *first* task of science, but only the *first*, is to obtain knowledge of this motion. But this mechanical motion does not exhaust motion as a whole. Motion is not merely change of place, in fields higher than mechanics it is also change of quality. The discovery that heat is a molecular motion was epoch-making. But if I have nothing more to say of heat than that it is a certain displacement of molecules, I should best be silent. Chemistry seems to be well on the way to

[1] See Appendix II, p. 330.

explaining a number of chemical and physical properties or elements from the ratio of the atomic volumes to the atomic weights. But no chemist would assert that all the properties of an element are exhaustively expressed by its position in the Lothar Meyer curve,[1] that it will ever be possible by this alone to explain, for instance, the peculiar constitution of carbon that makes it the essential bearer of organic life, or the necessity for phosphorus in the brain. Yet the " mechanical " conception amounts to nothing else. It explains all change from change of place, all qualitative differences from quantitative, and overlooks that the relation of quality and quantity is reciprocal, that quality can become transformed into quantity just as much as quantity into quality, that, in fact, reciprocal action takes place. If all differences and changes of quality are to be reduced to quantitative differences and changes, to mechanical displacement, then we inevitably arrive at the proposition that all matter consists of *identical*, smallest particles, and that all qualitative differences of the chemical elements of matter are caused by quantitative differences in number and by the spatial grouping of those smallest particles to form atoms. But we have not got so far yet.

It is our modern natural scientists' lack of acquaintance with any other philosophy than the most mediocre vulgar philosophy, like that now rampant in the German universities, which allows them to use expressions like " mechanical " in this way, without taking into account, or even suspecting, the consequences with which they thereby necessarily burden themselves. The theory of the absolute qualitative identity of matter has its supporters—empirically it is equally impossible to refute it or to prove it. But if one asks these people who want to explain everything " mechanically " whether they are conscious of this consequence and accept the identity of matter, what a variety of answers will be heard !

The most comical part about it is that to make " materialist " equivalent to " mechanical " derives from Hegel, who wanted to throw contempt on materialism by the addition " mechanical." Now the materialism criticised by Hegel—the French materialism of the eighteenth century

[1] In which atomic volumes are plotted against atomic weights.

—was in fact exclusively *mechanical*, and indeed for the very natural reason that at that time physics, chemistry, and biology were still in their infancy, and were very far from being able to offer the basis for a general outlook on nature. Similarly Hæckel takes from Hegel the translation : *causae efficientes*=mechanically acting causes, and *causae finales*=purposively acting causes ; where Hegel, therefore, puts mechanical as equivalent to blindly acting, unconsciously acting, and not as equivalent to mechanical in Hæckel's sense of the word.[1] But this whole antithesis is for Hegel himself so much a superseded standpoint that he *does not even mention it* in either of his two accounts of causality in his *Logic*—but only in his *History of Philosophy*, in the place where it comes historically (hence a sheer misunderstanding on Hæckel's part due to superficiality !) and quite incidentally in dealing with teleology (*Logic*, III, II, 3) where he mentions it as the form in which the *old metaphysics* conceived the antagonism of mechanism and teleology, but otherwise treating it as a long superseded standpoint.[2] Hence Hæckel copied incorrectly in his joy at finding a confirmation of his " mechanical " conception and so arrives at the beautiful result that if a particular change is produced in an animal or plant by natural selection it has been effected by a *causa efficiens*, but if the same change arises by *artificial* selection then it has been effected by a *causa finalis* ! The breeder as *causa finalis* ! Of course a dialectician of Hegel's calibre could not be caught in the vicious circle of the narrow opposition of *causa efficiens* and *causa finalis*. And for the modern standpoint the whole hopeless rubbish about this opposition is put an end to because we *know* from experience and from theory that both matter and its mode of existence, motion, are uncreatable and are, therefore, their own final cause ; while to give the name *effective* causes to the individual causes which momentarily and locally become isolated in the mutual interaction of the motion of the universe, or which are isolated by our reflecting mind, adds absolutely no new determination but only a confusing element. A cause that is not effective is no cause.

     N.B.—Matter as such is a pure creation of thought and an

[1,2] See Appendix II, p. 380.

abstraction. We leave out of account the qualitative difference of things in comprehending them as corporeally existing things under the concept matter. Hence matter as such, as distinct from definite existing pieces of matter, is not anything sensuously existing. If natural science directs its efforts to seeking out uniform matter as such, to reducing qualitative differences to merely quantitative differences in combining identical smallest particles, it would be doing the same thing as demanding to see fruit as such instead of cherries, pears, apples, or the mammal as such instead of cats, dogs, sheep, etc., gas as such, metal, stone, chemical compound as such, motion as such. The Darwinian theory demands such a primordial mammal, Hæckel's pro-mammal, but it, at the same time, has to admit that if this pro-mammal contains within itself in *germ* all future and existing mammals, it was in reality lower in rank than all existing mammals and exceedingly crude, hence more transitory than any of them. As Hegel has already shown, *Encyclopædia* I, p. 199,[1] this view is therefore " a one-sided mathematical standpoint," according to which matter must be looked upon as having only quantitative determination, but, qualitatively, as identical originally, "no other standpoint than that" of the French materialism of the eighteenth century. It is even a retreat to Pythagoras, who regarded number, quantitative determination as the essence of things.

In the first place, Kekulé. Then : the systematising of natural science, which is now becoming more and more necessary, cannot be found in any other way than in the interconnections of phenomena themselves. Thus the mechanical motion of small masses on any heavenly body ends in the contact of two bodies, which has two forms, distinct from one another only in degree, viz. friction and impact. So we investigate first of all the mechanical effect of friction and impact. But we find that they are not thereby exhausted : friction produces heat, light, and electricity, impact produces heat and light if not electricity also—hence conversion of motion of masses into molecular motion. We enter the realm of molecular motion, physics, and investigate further. But here too we find that molecular

[1] See Appendix II, p. 331.

motion does not represent the conclusion of the investigation. Electricity passes into and arises from chemical reaction. Heat and light, ditto. Molecular motion becomes transformed into motion of atoms—chemistry. The investigation of chemical processes is confronted by the organic world as a field for research, that is to say, a world in which chemical processes take place, although under different conditions, according to the same laws as in the inorganic world, for the explanation of which chemistry suffices. In the organic world, on the other hand, all chemical investigations lead back in the last resort to a body—protein—which, while being the result of ordinary chemical processes, is distinguished from all others by being a self-acting, permanent chemical process. If chemistry succeeds in preparing this protein, a so-called protoplasm, with the specific nature which it obviously had at its origin, a specificity, or rather absence of specificity, such that it contains potentially within itself all other forms of protein (though it is not necessary to assume that there is only one kind of protoplasm), then the dialectical transition has also been accomplished in reality, hence completely accomplished. Until then, it remains a matter of thought, alias of hypothesis. When chemistry produces protein, the chemical process will reach out beyond itself, as in the case of the mechanical process above, that is, it will come into a more comprehensive realm, that of the organism. Physiology is, of course, the chemistry and especially the physics of the living body, but with that it also ceases to be specially chemistry, on the one hand its domain becomes restricted but, on the other hand, inside this domain it becomes raised to a higher power.

(c) On Nägeli's Incapacity to Know the Infinite.

Nägeli,[1] pp. 12, 13.

Nägeli first of all says that we cannot know real qualitative differences, and immediately afterwards says that such " absolute differences " do not occur in Nature ! P. 12.

_____
[1] C. von Nägeli. *Über die Schranken der naturwissenschaftlichen Erkenntnis* [*The Limits of Scientific Knowledge*], September, 1877.

In the first place, every qualitative infinity has many quantitative gradations, *e.g.* shades of colour, hardness and softness, length of life, etc., and these, although qualitatively distinct, are measurable and knowable.

In the second place, qualities do not exist but only things *with* qualities and indeed with infinitely many qualities. Two different things always have certain qualities (properties attaching to corporeality at least) in common, others differing in degree, while still others may be entirely absent in one of them. If we consider two such extremely different things —*e.g.* a meteorite and a man—together but in separation, we get very little out of it, at most that heaviness and other corporeal properties are common to both. But an infinite series of other natural objects and natural processes can be put between the two things, permitting us to complete the series from meteorite to man and to allocate to each its place in the interconnection of nature and thus to *know* them. Nägeli himself admits this.

Thirdly, our various senses might give us absolutely different impressions as regards quality. According to this, properties which we experience by means of sight, hearing, smell, taste, and touch would be absolutely different. But even here the differences disappear with the progress of investigation. Smell and taste have long ago been recognised as allied senses belonging together, which perceive conjoint if not identical properties ; sight and hearing both perceive wave oscillations. The sense of touch and sight are mutually complementary to such an extent that from the appearance of an object we can often enough predict its tactile properties. And, finally, it is always the same " I " that receives and elaborates all these different sense impressions, that comprehends them into a unity, and likewise these various impressions are provided by the same thing, appearing as its *common* properties, and therefore helping us to know it. To explain these different properties, accessible only to different senses, to bring them into connection with one another, is therefore the task of science, which so far has not complained because we have not a general sense in place of the five special senses, or because we are not able to see or hear tastes and smells.

Wherever we look, nowhere in nature are there to be found

such " qualitatively or absolutely distinct fields," which are put forward as incomprehensible. The whole confusion springs from the confusion about quality and quantity. In accordance with the prevailing mechanical view, Nägeli regards all qualitative differences as explained only in so far as they can be reduced to quantitative differences (on which what is necessary to be said will be found elsewhere), or because quality and quantity are for him absolutely distinct categories. Metaphysics.

" We can know *only the finite*, etc." This is quite correct in so far as only finite objects enter the sphere of our knowledge. But the statement needs to be completed by this : " fundamentally we can know *only the infinite*." In fact all real, exhaustive knowledge consists solely in raising the single thing in thought from singularity into particularity and from this into universality in seeking and establishing the infinite in the finite, the eternal in the transitory. The form of universality, however, is the form of self-completeness, hence infinity ; it is the comprehension of the many finites in the infinite. We know that chlorine and hydrogen within certain limits of temperature and pressure and under the influence of light, combine with an explosion to form hydrochloric acid gas, and as soon as we know this, we know also that this *takes place everywhere* and *at all times* where the above conditions are present, and it can be a matter of indifference, whether this occurs once or is repeated a million times, or on how many heavenly bodies. The form of universality in nature is *law*, and no one talks more of *the eternal character of the laws of nature* than the natural scientist. Hence if Nägeli says that the finite is made impossible to establish by not desiring to investigate merely this finite, adding instead something eternal to it, then he denies either the possibility of knowing the laws of nature or their eternal character. All true knowledge of nature is knowledge of the eternal, the infinite, and hence essentially absolute.

But this absolute knowledge has an important drawback. Just as the infinity of knowable matter is composed of the purely finite, so the infinity of thought which knows the absolute is composed of an infinite number of finite human minds, working side by side and successively at this infinite knowledge, committing practical and theoretical blunders, setting out from erroneous, one-sided, and false premises, pursuing false, tortuous, and uncertain paths, and often not even finding the right one when they run their noses against it (Priestley[1]).

The cognition of the infinite is therefore beset with double difficulty and from its very nature can only take place in an infinite asymptotic progress. And that fully suffices us in order to be able to say : the infinite is just as much knowable as unknowable, and that is all that we need.

Curiously enough, Nägeli says the same thing : " We can know only the finite, but also we can know *all that is finite* that comes into the sphere of our sensuous perception." The finite that comes into the sphere, etc., constitutes in sum precisely the infinite, for *it is just from this that Nageli has derived his idea of the infinite !* Without this finite, etc., he would have indeed no idea of the infinite !

(Bad infinity, as such, to be dealt with elsewhere.)

(Before this investigation of infinity comes the following) :

(1) The " insignificant sphere " in regard to space and time.

(2) The " probably defective elaboration of the sense organs."

(3) That we can only know the finite, transitory, changing and what differs in degree, the relative, etc. (as far as), " we do not know what time, space, force and matter, motion and rest, cause and effect are."

It is the old story. First of all one makes sensuous things into abstractions and then one wants to know them through the senses, to see time and smell space. The empiricist becomes so steeped in the habit of empirical experience, that he believes that he is still in the field of sensuous knowledge when he is operating with abstractions. We know what an hour is, or a metre, but not what time and space are ! As if time was anything other than just hours, and space anything but just cubic metres ! The two forms of existence of matter are naturally nothing without matter, empty concepts, abstractions which exist only in our minds. But, of course, we are also not supposed to

----

[1] Priestley discovered oxygen without knowing it.

know what matter and motion are! Of course not, for matter as such and motion as such have not yet been seen or otherwise experienced by anyone, but only the various, actually existing material things and forms of motion. Matter is nothing but the totality of material things from which this concept is abstracted, and motion as such nothing but the totality of all sensuously perceptible forms of motion ; words like matter and motion are nothing but *abbreviations* in which we comprehend many different sensuously perceptible things according to their common properties. Hence matter and motion *cannot* be known in any other way than by investigation of the separate material things and forms of motion, and by knowing these, we also *pro tanto* know matter and motion *as such*. Consequently, in saying that we do not know what time, space, motion, cause, and effect are, Nägeli merely says that first of all we make abstractions of the real world through our minds, and then cannot know these self-made abstractions because they are creations of thought and not sensuous objects, while all knowing is *sensuous measurement!* This is just like the difficulty mentioned by Hegel, we can eat cherries and plums, but not *fruit*, because no one has so far eaten fruit as such.

When Nägeli asserts that there are probably a whole number of forms of motion in nature which we cannot perceive by our senses, that is a poor apology, equivalent to the suspension—at least for our knowledge—of the law of the uncreatability of motion. For they could certainly be *transformed into motion perceptible to us!* That would be an easy explanation of, for instance, contact electricity.

Ad vocem Nägeli. Impossibility of conceiving the infinite. As soon as we say that matter and motion are not created and are indestructible, we are saying that the world exists as infinite progress, *i.e.* in the form of bad infinity, and thereby we have conceived all of this process that is to be conceived. At the most the question still arises whether this process is an eternal repetition—in a great cycle—or whether the cycles have upward and downward portions.

APPENDIX II

SOURCE REFERENCES

In this appendix will be found extracts from various books and journals giving the text of the passages referred to by Engels in the course of his work. Where the source is an English one, the original English text is reproduced; where the work is a German or French one of which a standard English translation is available, this has been used. In other cases the passage has been translated from the original.

The page references are to the edition of which particulars are given in the Bibliography. Capital letters after the author's name refer to the particular volume listed under such letter in the Bibliography. Where two letters with page references are given, the first refers to the edition in the original language, the second to the English translation. The Notes to Appendix I precede those to the main text.

P. 319. *Nature*, 1877, XVII, p. 55 : " On entering upon the duties of rector of the University, Prof. Kekulé, the distinguished chemist, delivered, on October 18, a brilliant address on the scientific position of chemistry, and the fundamental principles of this science. He made the following definition of chemistry as distinct from physics and mechanics :—' Chemistry is the science of the statics and dynamics of atoms ; physics that of the statics and dynamics of molecules ; while mechanics consider the masses of matter consisting of a large number of molecules.' After rapidly sketching the growth of the present atomic theory, he claimed that the mass of results now obtained showed that chemistry was slowly but surely approaching its goal, the knowledge of the constitution of matter. In opposition to the opinion that theory should be banished from the exact sciences, he regarded it as an actual felt necessity of the human mind to classify the endless series of individual facts from general standpoints—at present of a hypothetical

nature—and that it was precisely the discussion of these
hypotheses which often led to the most valuable discoveries."

P. 320.   E. Hæckel, *Die Perigenesis der Plastidule*, D.,
p. 13 : " For if modern physiology quite rightly shuts the
door on vitalism and teleology, if it rejects all mystical and
supernatural action of a ' vital force ' kind, and in its field
allows only of the operation of physico-chemical—or, in
the wider sense, mechanical—forces, then it must also seek
such a mechanical explanation for the two most important
life activities in the development of form, viz. heredity, and
adaptation."

P. 322.   Hæckel, D., p. 13 : " And if our great critical
philosopher, *Immanuel Kant*, quite rightly demands of
natural science that it should put mechanical causes (*causae
efficientes*) everywhere in place of purposive causes (*causae
finales*) ; if *Kant* further asserts that mechanism alone
contains a real explanation of phenomena and that ' without
the principle of mechanism in nature there can be no natural
science,' we shall also recognise this monistie standpoint
as the only justifiable one for our history of evolution as a
true natural science, and may also seek only mechanical
causes for the physical facts of organic evolution."

P. 322.   Hegel, *Logik*, A., II, p. 209, B., II, p. 374 : " When
adequacy to an End is perceived, an understanding is
assumed as its origin : that is, the proper and free existence
of the Notion is demanded for the End. Teleology is
chiefly contrasted with Mechanism, where the determinateness posited in the Object, being external, is essentially of
such a kind as manifests no self-determination. The
opposition between *causae efficientes* and *causae finales*—
between merely efficient and final causes—refers to this
distinction ; and to this, taken in a concrete form, the
investigation reverts whether the absolute essence of the
world must be taken as a blind natural Mechanism or as an
understanding which determines itself by Ends. The
antinomy between fatalism or determinism and freedom
also regards the opposition between Mechanism and Teleology ; for the Free is the Notion in its existence."

P. 323. Hegel, *Enzyklopädie*, C., I, p. 199 : " Moreover on closer examination the exclusively mathematical standpoint mentioned here, which identifies quantity, this determinate stage of the logical Idea, with the latter itself, is no other than that of *materialism*, as indeed finds full confirmation in the history of scientific consciousness, especially in France, since the middle of the last century." (Translated from the German ; the English rendering, D., p. 187, is inadequate.)

P. 30. Hegel, *Logik*, A., I/1, p. 433. B., I, p. 376: "Chemical substances are the properest examples of such Measures, as, being Measure-moments, have that which constitutes their determination only in their attitude to others. Acids and alkalis or bases appear as things immediately determinate in themselves, but still more as incomplete elements of bodies, as constituent parts which do not really exist for themselves, but have this existence only, that they transcend their isolated persistence and combine with another. Further, this distinctness which makes them stable does not consist in this immediate quality, but in the quantitative character of the attitude. For this is not confined to the chemical opposition of acid and alkali or base in general, but is specified to be a measure of saturation, and consists in the specific determinateness of the quantity of the substances which neutralise one another. This quantity-determination in respect of saturation constitutes the qualitative nature of a substance : it makes it what it is for itself, and the number which expresses this is essentially one of several exponents for an opposed unit."

P. 49. Hegel, *Enzyklopädie*, C., I, pp. 272–3. D., pp. 249–50 : " But the different forces themselves are a multiplicity again, and in their mere juxtaposition seem to be contingent. Hence in empirical physics, we speak of the forces of gravity, magnetism, electricity, etc., and in empirical psychology of the forces of memory, imagination, will, and all the other faculties. All this multiplicity again excites a craving to know these different forces as a single whole, nor would this craving be appeased even if the several forces were traced back to one common primary force. Such a primary force would be really no more than an empty abstraction, with as little content as the abstract thing-in-itself."

P. 64. Thomson and Tait, *Treatise on Natural Philosophy*, A., p. 163 : " The *Vis Viva*, or Kinetic Energy, of a moving body is proportional to the mass and the square of the velocity, conjointly. If we adopt the same units of mass and velocity as before, there is particular advantage in defining kinetic energy as half the product of the mass and the square of its velocity."

P. 75. Clerk Maxwell, *Theory of Heat*, pp. 87–9 : " Suppose a body whose mass is *m* (*m* pounds or *m* grammes) to be moving in a certain direction with a velocity which we shall call *v*, and let a force, which we shall call *f*, be applied to the body in the direction of its motion. Let us consider the effect of this force acting on the body for a very small time *t*, during which the body moves through the space *s*, and at the end of which its velocity is *v'*.

To ascertain the magnitude of the force *f*, let us consider the momentum which it produces in the body, and the time during which the momentum is produced.

The momentum at the beginning of the time *t* was *mv*, and at the end of the time *t* it was *mv'* so that the momentum produced by the force *f* acting for the time *t* is $mv' - mv$.

But since forces are measured by the momentum produced in unit of time, the momentum produced by *f* in one unit of time is *f*, and the momentum produced by *f* in *t* units of time is *ft*. Since the two values are equal,

$$ft = m \ (v' - v).$$

This is one form of the fundamental equation of dynamics. If we define the impulse of a force as the average value of the force multiplied by the time during which it acts, then this equation may be expressed in words by saying that the impulse of a force is equal to the momentum produced by it.

We have next to find *s*, the space described by the body during the time *t*. If the velocity had been uniform, the space described would have been the product of the time by the velocity. When the velocity is not uniform, the time must be multiplied by the means or average velocity to get the space described. In both these cases in which average force or average velocity is mentioned the time is supposed to be subdivided into a number of equal parts, and the average is taken of the force of the velocity for all these divisions of the time. In the present case, in which the time considered is so small that the change of velocity is also small, the average velocity during the time t may be taken as the arithmetical mean of the velocities at the beginning and at the end of the time, or

$$\tfrac{1}{2}(v + v').$$

Hence the space described is

$$s = \tfrac{1}{2}(v + v')t.$$

This may be considered as a kinematical equation, since it depends on the nature of motion only, and not on that of the moving body.

If we multiply together these two equations we get

$$fts = \tfrac{1}{2}m(v'^2 - v^2)t,$$

and if we divide by t we find

$$fs = \tfrac{1}{2}mv'^2 - \tfrac{1}{2}mv^2.$$

Now fs is the work done by the force f acting on the body while it moves in the direction of f through a space s. If we also denote $\tfrac{1}{2}mv^2$, the mass of the body multiplied by half the square of its velocity, by the expression *the kinetic energy of the body*, then $\tfrac{1}{2}mv'^2$ will be the kinetic energy after the action of the force f through a space s."

P. 76. Clausius, *Über den zweiten Hauptsatz*, A., p. 2: " For this reason I have proposed to introduce besides work [German : *Arbeit*] also a second magnitude, which it is true likewise represents work but measured not according to that mechanical measure, but according to the measure of heat, hence expressed in such a way that the work which is equivalent to the unit of heat is taken as the unit of work. For work defined in this way I have proposed the name *Werk* (work)."

P. 78. Clausius, *Die mechanische Wärmetheorie*, B., p. 22 : " While in earlier times the view was almost universally held that heat is a special substance, which is present in bodies in greater or less quantity and thereby determines their higher or lower temperature, and which is emitted by bodies and then with immense velocity traverses empty space and also such spaces as are contained in ponderable masses, and so forms radiant heat, more recently the view has gained ground that heat is a form of motion. Then the heat inside bodies, which determines the temperature of the latter, is regarded as a movement of the ponderable atoms, a movement in which the ether inside the body can also participate, and radiant heat is looked upon as a vibratory movement of the ether."

P. 86. Faraday, *Experimental Researches*, I, pp. 447–8 : " The *spark* is consequent upon a discharge or lowering of the polarized inductive state of many dielectric particles, by a particular action of a few of the particles occupying a very small and limited space ; all the previously polarized particles returning to their first or normal condition in the inverse order in which they left it, and uniting their powers meanwhile to produce, or rather to continue, the discharge effect in the place where the subversion of force first occurred. My impression is, that the few particles situated where discharge occurs are not merely pushed apart, but assume a peculiar state, a highly exalted condition for the time, *i.e.* have thrown upon them all the surrounding forces in succession, and rising up to a proportionate intensity of condition, perhaps equal to that of chemically combining atoms, discharge the powers, possibly in the same manner as they do theirs, by some operation at present unknown to us ; and so the end of the whole. The ultimate effect is exactly as if a metallic wire had been put into the place of the discharging particles ; and it does not seem impossible that the principles of action in both cases may, hereafter, prove to be the same."

P. 86. Hegel, *Naturphilosophie*, E., p. 349 : " Electricity makes its appearance whenever two bodies touch one another, especially when they are rubbed. Hence electricity is not only to be found on the electrical machine ; on the contrary, every pressure, every blow, sets up electrical stress, but contact is the condition for this. Electricity is no specific special phenomenon which only occurs on amber, shellac, etc., on the contrary it occurs on every body that is in contact with another body ; it is only a matter of having a very delicate electrometer to convince oneself of this. The angry self of the body is exhibited by every body when excited ; all manifest this vitality towards one another."

P. 138. Naumann, *Handbuch der allgemeinen und physikalischen Chemie*, p. 729 : " Although, therefore, the water above-mentioned could be regarded as one of the purest ever prepared, it cannot be maintained that it was perfectly pure, and that the value $k = \cdot 000000000072$ is not to be regarded as an upper limit. In practice, it is true that even this figure gives water the significance of a galvanic non-conductor, for it can easily be calculated that a column of the above water 1 mm. in length offers the same resistance as a copper conductor of the same cross-section and of a length about equal to the diameter of the moon's orbit."

P. 152. Büchner, *Kraft und Stoff*, A., Preface, pp. vi–vii, B., Preface, pp. xvii–xviii : " The scholastic philosophy, still riding upon its high, though terribly emaciated, horse, conceived that it has long ago done with such theories, and has assigned them, ticketed ' materialism,' ' sensualism,' ' determinism,' to the scientific lumber-room, or, as the phrase goes, has assigned them their ' historical value.' But this philosophy sinks daily in the estimation of the public and loses its ground opposed to natural science, which gradually establishes the fact that macrocosmic and microcosmic existence obeys in its origin, life, and decay, *mechanical* laws inherent in the things themselves. . . .

Philosophical expositions which cannot be grasped by every educated person do not, in our opinion, deserve the printers ink expended on them. What has been clearly thought out can also be said clearly and without circumlocution. The philosophical evils which disfigure the writings of the erudite seem to aim more at concealing thoughts than at revealing them. The times of erudite heroics, of philosophical charlatanism, or ' intellectual legerdemain,' as *Cotta* very characteristically expresses it, are over or ought to be over. . . .

Proceeding from the fixed relation between matter and force as an indestructable basis, empirical philosophy must arrive at results which discard every kind of supernaturalism and idealism in the explanation of natural events, considering the latter as perfectly independent of any external power. The final victory of this kind of philosophical cognition cannot be doubted. The strength of its proof lies in facts, not in unintelligible and empty phrases. There is, in the end, no fighting against facts; it is like kicking against the pricks. It is needless to observe that our expositions have nought in common with the conceptions of the old " natural-philosophical " school. The singular attempts to construe nature out of thought instead of from observation have failed, and brought the adherents to that school into such discredit, that the name ' natural philosopher ' has become a byeword and a nickname."

P. 152. Hegel, *Geschichte der Philosophie*, F., III, pp. 529–30 : " The Germans, who honourably enough wanted to deal with the matter in a really thorough-going way, and who wanted to substitute the basis of reason for wit and vivacity, which indeed does not really demonstrate wit and vivacity, in this way obtained so empty a content that nothing can be more tedious than this thorough-going treatment; as in Eberhard, Tetens, etc.

Nicolai, Mendelssohn, Selzer and their like also philosophised chiefly about taste and the fine arts; for the Germans ought also to have fine literature and art. However, they only arrived thereby at the limit of meagreness in aesthetics—Lessing had called it shallow twaddle."

P. 153. Büchner, *Kraft und Stoff*, A., pp. 170–1 : " It follows from this, and is in the closest connection with it, that we can have no science, no notion of the *Absolute*, *i.e.* of that which goes beyond the perceptual world surrounding us. However much the metaphysicians may vainly exert themselves to define the Absolute, however much religion may strive to awaken faith in the absolute by acceptance of immediate revelation : nothing can cover this inner defect. All our knowledge and conceiving is relative and proceeds only from a mutual comparison of the perceptual things surrounding us. We would have no notion of dark without light, no idea of high without low, or warm without cold, and so on ; we do not possess any absolute ideas. We are not capable of constructing even a more remote idea of ' eternal ' or ' infinite,' because our understanding in its perceptual limitation by space and time finds impassable barrier to such an idea. Because in the perceptual world we are accustomed to find a cause everywhere where we see an effect, we have falsely inferred the existence of a supreme cause of all things, although such a cause cannot come within the domain of our other ideas as it is in conflict with scientific experience."

P. 155 (2). Hegel, *Enzyklopädie*, C., I, p. 9, D., p. 9 : " Everybody allows that to know any other science you must have first studied it, and that you can only claim to express a judgment upon it in virtue of such knowledge. Everybody allows that to make a shoe you must have learned and practised the craft of the shoemaker, though every man has a model in his own foot, and possesses in his hands the natural endowments for the operations re-quired. For philosophy alone, it seems to be imagined, such study, care, and application are not in the least requisite."

P. 155 (3). Hegel, *Enzyklopädie*, C., I, p. 11, D., p. 11 : " This divorce between idea and reality is especially dear to the analytic understanding which looks upon its own abstractions, dreams though they are, as something true and real, and prides itself on the imperative ' ought,' which it takes especial pleasure in prescribing even on the field of politics. As if the world had waited on it to learn how it ought to be, and was not ! For, if it were as it ought to be, what would come of the precocious wisdom of that ' ought ' ? "

P. 155 (4). Hegel, *Enzyklopädie*, C., I, p. 35, D., p. 36–7 : " For the explanation of *Sense*, the readiest method certainly is, to refer to its external source—the organs of sense. But to name the organ does not help much to explain what is apprehended by it. The real distinction between sense and thought lies in this—that the essential feature of the sensible is individuality, and as the individual (which, reduced to its simplest terms is the atom) is also a member of a group, sensible existence presents a number of mutually exclusive units—of units, to speak in more definite and abstract formulæ, which exist side by side with, and after, one another."

P. 155 (5). Hegel, *Enzyklopädie*, C., I, p. 40, D., p. 41–2 : " For instance, we observe thunder and lightning. The phenomenon is a familiar one, and we often perceive it. But man is not content with a bare acquaintance, or with the fact as it appears to the senses ; he would like to get behind the surface, to know what it is, and to comprehend it. This leads him to reflect : he seeks to find out the cause as something distinct from the mere phenomenon : he tries to know the inside in its distinction from the outside. Hence the phenomenon becomes double, it splits into inside and outside, into force and its manifestation, into cause and effect. Once more we find the inside or the force identified with the universal and permanent : not this or that flash of lightning, this or that plant—but that which continues the same in them all. The sensible appearance is individual and evanescent : the permanent in it is discovered by reflection. Nature shows us a countless number of individual forms and phenomena. Into this variety we feel a need of introducing unity : we compare, consequently, and try to find the universal of each single case. Individuals are born and perish : the species abides and recurs in them all : and its existence is only visible to reflection. Under the same head fall such laws as those regulating the motion of the heavenly bodies. To-day we see the stars here, and tomorrow there : and our mind finds something incongruous in this chaos— something in which it can put no faith, because it believes in order and in a simple, constant, and universal law. Inspired by this belief, the mind has directed its reflection towards the phenomena, and learnt their laws. In other words, it has established the movement of the heavenly bodies to be in accordance with a universal law from which every change of position may be known and predicted."

P. 155 (6). Hegel, *Enzyklopädie*, C., I, p. 42, D., p. 43 :
" What reflection elicits, is a product of our thought. Solon, for instance, produced out of his head the laws he gave to the Athenians."

P. 155 (7). Hegel, *Enzyklopädie*, C., I, p. 45, Ð., p. 45 :
" *Logic therefore coincides with Metaphysics, the science of things set and held in thoughts*—thoughts accredited able to express the essential reality of things."

P. 155 (8). Hegel, *Enzyklopädie*, I, p. 53, D., pp. 52-3 :
" For in experience everything depends upon the mind we bring to bear upon actuality. A great mind is great in its experience ; and in the motley play of phenomena at once perceives the point of real significance."

P. 159 (1). A. von Haller's poem, " *Die menschlichen Tugenden* " (*The human virtues*), which appeared in 1732, was first contradicted by Goethe in 1820 in his poem " *Allerdings* " and later in his poem " *Ultimatum* " :

" Ins Innre der Natur dringt kein erschaffener Geist
Zu glücklich, wenn er nur die äussere Schale weist "
Das hör 'ich sechzig Jahre wiederholen,
Und fluche darauf, aber verstohlen,—
Natur hat weder Kern noch Schale
Alles est sie mit einem Male.

(" No mortal mind can Nature's inner secrets tell
Too happy only if he knows the outer shell "
For sixty years to this I've had to hark,
I curse the sentiment, but keep it dark,—
Nor shell nor kernel Nature does possess
Is everything at once and nothing less.)

P. 159 (2). Hegel, *Enzyklopädie*, C., I, p. 95, D., pp. 91-2 :
" It follows that the categories are not fit terms to express the Absolute—the Absolute not being given in perception :
—and Understanding, or knowledge by means of the categories, is consequently incapable of knowing the Things-in-themselves.

The Thing-in-itself (and under ' thing ' is embraced
even Mind and God) expresses the object when we leave
out of sight all that consciousness makes of it, all its emotional
aspects, and all specific thoughts of it. It is easy to see
what is left—utter abstraction, total emptiness, only
described still as an ' other-world '—the negative of every
image, feeling, and definite thought. Nor does it require
much penetration to see that this *caput mortuum* is still only
a product of thought, such as accrues when thought is
carried on to abstraction, unalloyed : that it is the work
of the empty ' Ego,' which makes an object out of this
empty self-identity of its own. The *negative* characteristic
which this abstract identity receives as an *object*, is also
enumerated among the categories of Kant, and is no less
familiar than the empty identity aforesaid. Hence one
can only read with surprise the perpetual remark that we
do not know the Thing-in-itself. On the contrary, there is
nothing we can know so easily."

P. 161. Hegel, *Enzyklopädie*, C., I, p. 222, D., p. 206 :
" In the sphere of Being the reference of one term to another,
is only implicit ; in Essence on the contrary it is explicit.
And this in general is the distinction between the forms of
Being and Essence : in Being everything is immediate, in
Essence everything is relative."

P. 162. Hegel, *Enzyklopädie*, C., I, p. 268, D., pp. 245–6 :
" The relation of whole and parts, being the immediate
relation, comes easy to reflective understanding ; and for
that reason it often satisfies when the question really turns
on profounder ties. The limbs and organs, for instance,
of an organic body are not merely parts of it : it is only
in their unity that they are what they are, and they are
unquestionably affected by that unity, as they also in turn
affect it. These limbs and organs become mere parts,
only when they pass under the hands of the anatomist,
whose occupation, be it remembered, is not with the living
body but with the corpse. Not that such analysis is illegitimate : we only mean that the external and mechanical
relation of whole and parts is not sufficient for us, if we want
to study organic life in its truth."

P. 163 (1).   Hegel, *Enzyklopädie*, C., I, p. 235, D., p. 217 :
" We do not stop at this point however, or regard things
merely as different.   We compare them one with another,
and thus discover the features of likeness and unlikeness.
The work of the finite sciences lies to a great extent in the
application of these categories, and the phrase ' scientific
treatment ' generally means no more than the method
which has for its aim comparison of the objects under
examination."

P. 163 (2).   Hegel, *Enzyklopädie*, C., I, p. 231, D., p. 214 :
" It is important to come to a proper understanding on
the true meaning of Identity : and, for that purpose, we
must especially guard against taking it as abstract Identity,
to the exclusion of all Difference.   That is the touch-stone
for distinguishing all bad philosophy from what alone deserves
the name of philosophy."

P. 164.   Hegel, *Enzyklopädie*, C., I, p. 152, D., p. 148 :
" We say, for instance, that man is mortal, and seem to
think that the ground of his death is in external circumstances
only ; so that if this way of looking were correct, man
would have two special properties, vitality and—also—
mortality.   But the true view of the matter is that life, as
life, involves the germ of death, and that the finite, being
radically self-contradictory, involves its own self-sup-
pression."

P. 165 (1).   Hegel, *Enzyklopädie*, C., I, p. 256, D., p. 235 :
" An animal may be said to consist of bones, muscles, nerves,
etc. : but evidently we are here using the term ' consist '
in a very different sense from its use when we spoke of the
piece of granite as consisting of the above-mentioned
elements.   The elements of granite are utterly indifferent
to their combination : they could subsist as well without it.
The different parts and members of an organic body on the
contrary subsist only in their union : they cease to exist as
such, when they are separated from each other."

P. 165 (2).   Hegel, *Enzyklopädie*, C., I, pp. 259–60, D.,
p. 238 : " The negation of the several matters, which is

insisted on in the thing no less than their independent existence, occurs in Physics as *porosity*. Each of the several matters (colouring matter, odorific matter, and if we believe some people, even sound-matter—not excluding caloric, electric matter, etc.) is also negated : and in this negation of theirs, or as interpenetrating their pores, we find the numerous other independent matters, which, being similarly porous, make room in turn for the existence of the rest. Pores are not empirical facts ; they are figments of the understanding, which uses them to represent the element of negation in independent matters. The further working-out of the contradictions is concealed by the nebulous imbroglio in which all matters are independent and all no less negated in each other. If the faculties or activities are similarly hypostatised in the mind, their living unity similarly turns to the imbroglio of an action of the one on the others.

These pores (meaning thereby not the pores in an organic body, such as the pores of wood or of the skin, but those in the so-called ' matters,' such as colouring matter, caloric, or metals, crystals, etc.) cannot be verified by observation. In the same way matter itself—furthermore form which is separated from matter—whether that be the thing as consisting of matters, or the view that the thing itself subsists and only has proper ties—is all a product of the reflective understanding which, while it observes and professes to record only what it observes, is rather creating a metaphysic, bristling with contradictions of which it is unconscious."

P. 168.   Hegel, *Naturphilosophie*, E., p. 79 : " But the point is not that such a tendency *exists*, but that it exists for itself separate from gravitation, as conceived in a completely independent form in *force*.   In the same place, Newton assures us that a lead bullet *in coelos abiret et motu abeundi pergeret in infinitum*, if (certainly if) *only* the appropriate velocity could be imparted to it.   Such separation of external and essential motion belongs neither to experience nor to the notion but only to abstracting reflection.   It is one thing to *distinguish* them, as is necessary, as well as to characterise them mathematically as separate lines, to treat them as separate quantitative factors, and so on—it is another thing to regard them as physically independent existences."

P. 169 (1). Hegel, *Naturphilosophie*, E., p. 65 : " Its essence [of motion] is to be the immediate unity of space and time : it is time really persisting through space, or space which is only made truly distinct through time. Thus we know that space and time belong to motion ; the velocity, the quantum of motion is space in relation to a definite time that has elapsed. One says also, motion is the relation of space and time ; the deeper manner of this relation, however, remained to be grasped. Only in motion have space and time reality."

P. 169 (2). Hegel, *Naturphilosophie*, E., p. 67 : " Space and time are filled with matter. Space is not conformable to its notion ; hence it is the concept of space itself that creates its existence in matter. Often a beginning has been made with matter, and then space and time regarded as forms of matter. What is correct is that matter is the real in regard to space and time. But the latter, on account of their abstraction, have to appear to us here as the primary ; and then it must be shown that matter is their truth. Just as there is no motion without matter, so also there is no matter without motion. Motion is the process, the transition from time into space and *vice versa* : matter, on the other hand, the relation of space and time, as latent identity. Matter is the primary reality, the existing Being-for-itself ; it is not only abstract being, but positive persistence of space, as excluding, however, other space."

P. 174. Grove, *The Correlation of Physical Forces*, pp. 10–14 : " Instead of regarding the proper object of physical science as a search after essential causes, I believe it ought to be, and must be, a search after facts and relations —that although the word Cause may be used in a secondary and concrete sense, as meaning antecedent forces, yet in an abstract sense it is totally inapplicable : we cannot predicate of any physical agency that it is abstractedly the cause of another ; and if, for the sake of convenience, the language of secondary causation be permissible, it should be only with reference to the special phenomena referred to, as it can never be generalised.

The misuse, or rather varied use, of the term Cause, has been a source of great confusion in physical theories, and philosophers are even now by no means agreed as to their conception of causation. The most generally received view of causation, that of Hume, refers it to invariable ante-cedence—*i.e.* we call that a cause which invariably precedes, that an effect which invariably succeeds. Many instances of invariable sequence might however be selected, which do not present the relations of cause and effect, thus, as Reed observes and Brown does not satisfactorily answer, day invariably precedes night, and yet day is not the cause of night. The seed, again, precedes the plant, but is not the cause of it; so that when we study physical phenomena it becomes difficult to separate the idea of causation from that of force, and these have been regarded as identical by some philosophers. To take an example which will contrast these two views : if a floodgate be raised, the water flows out; in ordinary parlance, the water is said to flow because the floodgate is raised : the sequence is invariable ; no floodgate, properly so called, can be raised without the water flowing out, and yet in another, and perhaps more strict sense, it is the gravitation of the water which causes it to flow. But, though we may truly say that, in this instance, gravitation causes the water to flow, we cannot in truth abstract the proposition, and say, generally, that gravitation is the cause of water flowing, as water may flow from other causes, gaseous elasticity, for instance, which will cause water to flow from a receiver full of air into one that is exhausted ; gravitation may also, under certain circumstances, arrest instead of cause the flow of water.

Upon neither view, however, can we get at anything like abstract causation. If we regard causation as invariable sequence, we can find no case in which a given antecedent is the only antecedent to a given sequent : thus, if water could flow from no other cause than the withdrawal of a flood-gate, we might say abstractedly that this was the cause of water flowing. If, again, adopting the view which looks to causation as a force, we could say that water could be caused to flow only by gravitation, we might say abstractedly that gravitation was the cause of water flowing—but this we cannot say ; and if we seek and examine any other example, we shall find that causation is only predicable of it in the particular case, and cannot be supported as an abstract proposition ; yet this is constantly attempted. Nevertheless, in each particular case, where we speak of Cause, we habitually refer to some antecedent power or force : we never see motion or any change in matter take effect without regarding it as produced by some previous change ; and, when we cannot trace it to its antecedent, we mentally refer it to one ; but whether this habit be philosophically correct is by no means clear. In other words, it seems questionable, not only whether cause and effect are convertible terms with antecedence and sequence, but whether in fact cause does precede effect, whether force does precede the change in matter of which it is said to be the cause.

The actual priority of cause to effect has been doubted, and their simultaneity argued with much ability. As an instance of this argument it may be said, the attraction which causes iron to approach the magnet is simultaneous with, and even accompanies the movement of the iron ; the movement is evidence of the co-existing cause of force, but there is no evidence of any interval in time between the one and the other. On this view time would cease to be a necessary element in causation ; the idea of cause, except perhaps as referred to a primeval creation, would cease to exist ; and the same arguments which apply to the simultaneity of cause with effect would apply to the simultaneity of Force with Motion. We could not, however, even if we adopted this view, dispense with the element of time in the sequence of phenomena ; the effect being thus regarded as ever accompanied simultaneously by its appropriate cause, we should still refer it to some antecendent effect ; and our reasoning as applied to the successive production of all natural changes would be the same.

Habit and the identification of thoughts with phenomena so compel the use of recognised terms, that we cannot avoid the use of the word cause even in the sense to which objection is taken ; and if we struck it out of our vocabulary, our language, in speaking of successive changes, would be unintelligible to the present generation. The common error, if I am right in supposing it to be such, consists in the abstraction of cause, and in supposing in each case a general secondary cause—a something which is not the first cause, but which, if we examine it carefully, must have all the attributes of a first cause, and an existence independent of, and dominant over, matter."

P. 174.   Grove, *loc. cit.*, p. 20 : " In placing the weight on the glass, we have moved the former to an extent equivalent to that which it would again describe if the resistance were removed, and this motion of the mass becomes an exponent or measure of the force exerted on the glass ; while this is in the state of tension, the force is ever existing, capable of reproducing the original motion, and while in a state of abeyance as to actual motion, it is really acting on the glass.   The motion is suspended, but the force is not annihilated."

P. 174.   Grove, *loc. cit.*, p. 16 : " The term force, although used in very different senses by different authors, in its limited sense may be defined as that which produces or resists motion.   Although strongly inclined to believe that the other affections of matter, which I have above named, are, and will ultimately be resolved into, modes of motion, many arguments for which will be given in subsequent parts of this Essay, it would be going too far, at present, to assume their identity with it ; I therefore use the term force, in reference to them, as meaning that active principle inseparable from matter which is supposed to induce its various changes."

P. 176.   Hæckel. B., pp. 59–60, C., I, p. 65 : " We see then, according to Agassiz's conception, that the Creator, in producing organic forms, goes to work exactly as a human architect, who has taken upon himself the task of devising and producing as many different buildings as possible, for the most manifold purposes, in the most different styles, in various degrees of simplicity, splendour, greatness, and perfection.   This architect would perhaps at first choose four different styles for all these buildings, say the Gothic, Byzantine, Moorish, and Chinese styles. In each of these styles he would build a number of churches. palaces, garrisons, prisons, and dwelling-houses. Each of these different buildings he would execute in ruder and more perfect, in greater and smaller, in simpler and grander fashion, etc. However, the human architect would perhaps, in this respect, be better off than the divine Creator, as he would have perfect liberty in the number of graduated subordinate groups. The Creator, however, according to Agassiz, can only move within six groups or categories : the species, genus, family, order, class, and type. More than these six categories do not exist for him."

P. 176. Hæckel, *Schöpfungsgèschichte*, B., pp. 75–7, C., I., p. 84 : " But Goethe did not merely endeavour to search for such far-reaching laws, he also occupied himself most actively for a long time with numerous individual researches, especially in comparative anatomy. Among these, none is perhaps more interesting than the discovery of the *mid-jawbone in man*. As this is, in several respects, important to the theory of development I shall briefly explain it here. There exist in all mammals two little bones in the upper jaw, which meet in the centre of the face, below the nose, and which lie between the two halves of the real upper-jawbone. These two bones, which hold the four upper cutting-teeth, are recognised without much difficulty in most mammals ; in man, however, they were at that time unknown, and celebrated comparative anatomists even laid great stress upon the want of a mid-jawbone, as they considered it to constitute the principal difference between man and apes—the want of a mid-jawbone was, curiously enough, looked upon as the most human of all human characteristics. But Goethe could not accept the notion that man, who in all other corporeal respects was clearly only a mammal of higher development, should lack this mid-jawbone. By the general law of induction as to the mid-jawbone he arrived at the special deductive conclusion that it must exist in man also, and Goethe did not rest until, after comparing a great number of human skulls, he really found the mid-jawbone. In some individuals it is preserved throughout the whole lifetime, but usually at an early age it coalesces

with the neighbouring upper jawbone and is therefore only
to be found as an independent bone in youthful skulls.   In
human embryos it can now be pointed out at any moment.
In man, therefore, the mid-jawbone actually exists and to
Goethe the honour is due of having first firmly established
this fact, so important in many respects ; and this he did
while opposed by the celebrated anatomist, Peter Camper,
one of the most important professional authorities.   The
way by which Goethe succeeded in establishing this fact is
especially interesting ; it is the way by which we continually
advance in biological science, namely, by way of induction
and deduction.   *Induction* is the inference of a general law
from the observation of numerous individual cases ; *deduction*,
on the other hand, is an inference from this general law
applied to a single case which has not yet been actually
observed.   From the collected empirical knowledge of those
days, the inductive conclusion was arrived at that all
mammals have had mid-jawbones.   Goethe drew from this
the deductive conclusion, that man, whose organisation
was in all other respects not essentially different from
mammals, must also possess this mid-jawbone ; and on close
examination it was actually found.   The deductive conclusion was confirmed and verified by experience."

P. 176.   Hæckel, *Schöpfungsgeschichte*, B., pp. 89–90,
C., I, p. 84 : " The teleological view of nature, which
explains the phenomena of the organic world by the action
of a personal Creator acting for a definite purpose, necessarily
leads, when carried to its extreme consequences, either to
utterly untenable contradictions, or to a two-fold (dualistic)
conception of nature, which most directly contradicts the
unity and simplicity of the supreme laws which are everywhere perceptible.   The philosophers who embrace teleology must necessarily assume two fundamentally different
natures : an *inorganic* nature, which must be explained by
causes acting mechanically (*causae efficientes*) ;   and an
organic nature, which must be explained by *causes acting for
a definite purpose* (*causae finales*)."

P. 176.   Hegel, *Geschichte der Philosophie*, F., III,
pp. 603–4 : " In this connection, Kant arrives at the

following: We would find no difference between the *mechanism of nature* and the *technique of nature, i.e.* the linking of End in nature, if our understanding were not of such a kind that it has to go from the universal to the particular, and that therefore the power of judgement cannot pronounce any determinate judgements without having a general law under which the particular could be subsumed. The particular as such, however, in respect to the universal contains something accidental, but reason equally demands that in the connection of the particular laws of nature there should also be unity, and indeed obedience to law, such obedience to law on the part of the accidental being called adequacy to an End : and the derivation of the particular laws from the general laws is, in respect of the accidental that the former contain, impossible *a priori* by determination of the notion of the object ; the notion of the adequacy to an End of nature in its products thus becomes a notion necessary for the human power of judgement, but one that does not concern the determination of the objects themselves, hence a subjective principle, and also only a leading thought for the power of judgement, whereby nothing can be said of its Being-in-itself."

P. 177. *Nature,* X, pp. 309–19 : Inaugural address of Prof. John Tyndall, at the forty-fourth annual meeting of the British Association for the Advancement of Science.

P. 180. Hæckel, *Schöpfungsgeschichte,* B., p. 382, C., II, p. 56 : " As some of these cells at an early stage encased themselves by secreting a hardened membrane, they formed the first vegetable cells, while others, remaining naked, developed into the first aggregates of animal cells.

The presence or absence of an encircling hard membrane forms the most important, although by no means the entire difference of form between animal and vegetable cells."

P. 181. *Ibid.,* B., pp. 384–5, C., II, p. 59 : " *Labyrinthläufer (Labyrinthuleae).* They are spindle-shaped cells, mostly of a yellow-ochre colour, which are sometimes united into a dense mass, sometimes move about in a very peculiar way. They form, in a manner not yet explained, a retiform frame of tangled threads (compared to a labyrinth) and on the dense filamentous ' tramways ' of this frame they glide about. From the shape of the cells of the *Labyrinthuleae* we might consider them as the simplest plants, from their motion, as the simplest animals, but in reality they are neither animals or plants."

P. 181. *Ibid*, B., pp. 409–10, C., II, pp. 87–8 : " Even the *single-celled primary plants*—which are distinguished from the monocytods (monocytoden) by possession of a kernel (nucleus), develop into a great variety of exquisite forms by adaptation ; this is the case especially with the beautiful *Desmidiaceae*. . . . It is very probable that similar primeval plants, the soft body of which, however, was not capable of being preserved in the fossil state, at one time peopled the Laurentian primeval sea in great masses and varieties, and in a great abundance of forms without, however, going beyond the stage of individuality of a single plastid."

P. 181. *Ibid*, B., p. 384, C., II, p. 59 : " This wonderful organism, which sometimes appears like a simple Amoeba, sometimes as a single fringed cell, sometimes as a many-celled fringed ball, can evidently be classed with none of the other Protista and must be regarded as the representative of a new independent group. As this group stands midway between several Protista and links them together, it may bear the name *Mediator*, or *Catallacta*."

P. 181. *Ibid*, B., pp. 383–4, C., II, pp. 58–9 : " A very remarkable new form of Protista, which I have named Flimmer ball (*Magosphaera*) I discovered in September 1869 on the Norwegian coast, and have more accurately described in my *Biological Studies*. Off the island of Gis-oe, near Bergen, I found swimming about, on the surface of the sea, extremely neat little balls composed of a number (about 30–40) of fringed pear-shaped cells, the pointed ends of which were united in the centre like radii. After a time the ball dissolved. The individual cells swarmed about independently in the water like fringed Infusoria or Ciliates. These afterwards sank to the bottom, drew their fringes into their bodies, and gradually changed into the form of creeping Amoebae. These last afterwards encased themselves and then divided by repeated halvings into a large number of cells. The cells became covered with vibratile hairs, broke through the case enclosing them, and now again swam about in the shape of a fringed ball."

P. 181. *Ibid*, B., p. 452 : " All other animals, from the simplest plant-animals to the vertibrates, from the sponge up to man, are composed of various kinds of tissues and organs, which all develop originally from two distinct cell layers. These two layers are the two primary germinal sheaths, with which we have already become acquainted in the embryonic developmental form of the gastrula. The outer cell layer, or the animal germinal sheath (the main sheath or Exoderma) is the basis for the animal organs of the animal body : skin, nervous system, muscular system, skeleton, etc. The inner cell layer, or the vegetative germinal sheath (the gut sheath or Entoderma), on the other hand, provides the material for the vegetative organs : digestive tract, vascular system, etc. Even to-day among the lower representatives of all six higher branches of animals we encounter the gastrula in the history of the embryo, where these two primary germinal sheaths occur in their simplest form and enclose the oldest primitive organ, the primordial gut with the primordial mouth. Hence we can group all these animals (in contrast to the primordial animals not equipped with a gut) as gut-possessing animals (Metazoa). All these gut-possessing animals can be derived from a common basic stock—*Gastraea*, and this long extinct basic stock must have been essentially similar in structure to the primitive embryonic form—the gastrula—which still occurs everywhere to-day. From this *Gastraea* there developed, as already shown, two distinct basic stocks, *Protascus* and *Prothelmis*, of which the former is to be regarded as the basic stock of the plant-animals, and the latter as the basic stock of the worms."

P. 182. Clausius, *Über den zweiten Hauptsatz*, A., pp. 6–7 : " Suppose one takes a quantity of a perfect gas, which occupies a definite volume. If this gas expands to another volume, for instance to twice the volume, then an increase

of disaggregation takes place which is completely determined
by the volume at the start and at the end.   At the same time,
on expansion heat becomes converted into work.   Since in a
perfect gas no internal work is performed, because the
molecules are already so far apart that their mutual influence
can be neglected, we have only to deal with the external
work which is performed by overcoming the external
pressure, hence with work the magnitude of which can be
easily stated.   The heat consumed for this work must be
communicated to the gas from outside, if its temperature is
to remain constant."

P. 189.   Hæckel, *Anthropogenie*, A., pp. 707–8 :   " According to the materialist view of the world, matter or substance
was present earlier than motion or vital force ;   *matter
created force*.   According to the spiritualist view of the
world, it is the other way round :   vital force or motion was
present earlier than matter, the latter being called into
being by it; *force created matter*.   Both views are dualistic
and we hold that both are equally false.   The opposition
between the two views is removed for us in the *monistic*
philosophy, which can as little conceive force without matter
as matter without force."   (Engels' italics.)

P. 201 (2).   Mädler, *Der Wunderbau des Weltalls*, pp. 42–3 :
" If it is . . . probable that the spaces between the heavenly
bodies are not absolutely empty but filled with a so-called
resistant medium (the ether), the latter, however, shows
not a single one of the properties which are essential attributes
of air and every form of gas. . . ."

P. 201 (2).   Lavrov, *Opyt istorii mysli*, pp. 103-4: "Immeasurable space, with its scattered worlds, is empirically
for us *empty* space, in which occur as isolated islands masses
of ponderable substance, which gravitate towards one another
and move under influence of this gravitation, which thus
constitutes the most general cosmic phenomenon accessible to
us.

But even this ponderable gravitating world is accessible
to us only as regards a very minute fraction, namely, as
regards the space of which we obtain knowledge through

light phenomena. Many scientists are of the opinion that these light phenomena cease for us when the so-called light wave has traversed a certain definite distance ; light is extinguished and there are no means by which it is possible for us to convince ourselves of the existence of worlds of one kind or another beyond the bounds of this distance. We scarcely know even one island universe in immeasurable space, viz. the one to which we ourselves belong. By means of thought we can convince ourselves of the *probability* of the existence of other island universes beyond its boundaries, of the reality of which mankind will never be certain ; but everything that we *know* of the *universe* is restricted to *our* single island universe."

P. 201. Lavrov, p. 109 : " Dead suns with their dead systems of planets and satellites continue their motion in space as long as they do not fall into a new nebula in process of formation. Then the remains of the dead world become material for hastening the process of formation of the new world. The enormous glowing heat of the mass in which the new world is being prepared, rapidly melts and gasifies the dead world, but thereby the new world rapidly passes through one of the primary phases of its existence, in order to pass over to the subsequent phases and in due course for innumerable millenia to move in the form of a dark dead system, until in turn it succeeds in serving as material for a still newer world that is coming into being.

Such is the probable fate of all that exists in the universe. While some systems are living through the last million centuries of their existence, others have hardly passed through the first milliards of millenia of their separate being. Here a world long since dead obtains the possibility of entering on the process of formation of a new solar system, there a world in formation which has come close to rigid masses is disintegrated into comets and falling stars. Violent death threatens worlds just as easily as inevitable natural extinction. But eternal motion does not cease, and new worlds eternally develop in place of former ones."

P. 203. Hæckel, *Schöpfungsgeschichte*, B., pp. 165–6 : " But perhaps the most remarkable of all Monera was discovered in 1868 by the famous English zoologist Huxley and named by him *Bathybius Haeckelii*. 'Bathybius' means : 'living in the depths.' For this wonderful organism lives in the enormous abysses of the ocean, which have become known to us in the last decade owing to painstaking British investigations, and which reach a depth of more than 12,000 feet, indeed in many places more than 24,000 feet.   Here Bathybius occurs in masses between the numerous Polythalamias and Radiolarias that populate the fine chalk-like mud of these abysses.   It is found partly in the form of rounded or formless granules of slime, partly in the form of a loose slime network covering bits of stone and other objects.   Small calcareous particles (Discoliths, Cynatholiths, etc.) are often embedded in these slimy gelatinous masses, probably excretory products of the latter. The whole body of this remarkable organism, Bathybius, consists, as in the case of other Monera, solely of structureless plasm or protoplasm, *i.e.* of the same protein-like carbon compound which occurs in infinite modifications in all organisms as the most essential and never absent bearer of life phenomena.   A detailed description and drawing of Bathybius and of other monera has been given by me in 1870 in my *Monograph of the Monera*."

P. 205.   Kopp, *Die Entwickelung der Chemie in der neueren Zeit*, p. 105 : " As regards the theory of affinity, the period now under consideration had inherited all the knowledge that was immediately at the disposal of the new era beginning with *Lavoisier*.   It is true that among some persons recollections of earlier false theories persisted, especially of the theory that the possibility of two bodies combining indicated some common content of both in regard to a given constituent, or in other words that the combining ability of bodies rested on kinship between them in the proper sense of the word.   But by most persons the concept of chemical attraction was already more correctly grasped."

P. 210.   Clerk Maxwell, *Theory of Heat*, p. 14 : " The distinguishing characteristic of radiant heat is that it travels in rays like light, whence the name radiant.   These rays have all the physical properties of rays of light and are capable of reflexion, refraction, interference, and polarisation.

They may be divided into different kinds by the prism as light is divided into its component colours, and some of the heat-rays are identical with the rays of light, while other kinds of heat-rays make no impression on our eyes."

P. 216. Clausius, *Über den zweiten Hauptsatz*, p. 16 : " One form of energy can be converted into another form of energy, but the quantity of energy never loses anything thereby, on the contrary the total energy present in the world is just as constant as the total amount of matter present in the world."

P. 226. Hegel, *Logik*, A., II, p. 154, B., II, p. 329 : " Rather, Induction is still essentially a subjective Syllogism. The middle consists of the individuals in their immediacy ; their comprehension into the genus by means of allness is an external reflection. Because of the persistent immediacy of the individuals and the externality which follows thence, universality is no more than completeness, or rather remains a problem to be solved. Consequently the progress to bad infinity once more emerges in universality : individuality is to be posited as identical with universality ; but, the individuals being equally posited as immediate, this unity remains no more than an enduring Ought ; it is a unity of equality ; the terms which are to be identical at the same time are not to be identical. a, b, c, d, e, but only to infinity, constitute the genus and give perfected experience. In so far the conclusion of Induction remains problematical."

P. 227. Hæckel, *Schöpfungsgeschichte*, B., p. 34, C., I, p. 37 : " In order, then, to avoid in future the usual confusion of this utterly objectionable Moral Materialism with our Scientific Materialism, we think it necessary to call the latter either *Monism* or *Realism*. The principle of this *Monism* is the same as what Kant terms ' the principle of mechanism,' and of which he expressly asserts, that *without it there can be no natural science at all.* This principle is quite inseparable from our Non-miraculous History of Creation, and characterises it as opposed to the teleological belief in the miracles of a Supernatural History of Creation."

P. 227.   Hæckel, B., pp. 90–1, C., I, p. 102 : " If we read Kant's *Criticism of the Teleological Faculty of Judgment*, his most important biological work, we perceive that in contemplating organic nature he always maintains what is essentially the teleological or dualistic point of view ; whilst for inorganic nature he unconditionally, and without reserve, assumes the mechanical or monistic method of explanation. He affirms that in the domain of inorganic nature all phenomena can be explained by mechanical causes, by the moving forces of matter itself, but not so in the domain of organic nature. In the whole of anorganology, in geology, mineralogy, in meteorology and astronomy, in the physics and chemistry of inorganic natural bodies, all phenomena are said to be explicable merely by mechanism (*causa efficiens*) without the intervention of a final purpose. In the whole domain of biology on the other hand—in botany, zoology, and anthropology—mechanism is not considered sufficient to explain to us all their phenomena ; but we are supposed to be able to comprehend them only by the assumption of a *final cause* acting for a definite purpose (*causa finalis*). In several passages Kant emphatically remarks that, from a strictly scientific point of view, *all* phenomena, without exception, require a mechanical interpretation *and that mechanism alone can offer a true explanation*. But at the same time he thinks, that in regard to living natural bodies, animals, and plants, our human power of comprehension is limited, and not sufficient for arriving at the real cause of organic processes, especially at the origin of organic forms. The *right* of human reason to explain all phenomena mechanically is unlimited, he says, but its *power* is limited by the fact that organic nature can be conceived only from a teleological point of view."

P. 228.   Hegel, *Logik*, A., II, pp. 247–8, B., II, p. 404 : " But simple Life not only is omnipresent, but is just the persistence and immanent substance of its objectivity ; as subjective substance, however, it is impulse—the specific

APPENDICES 357

impulse of the particular distinction, while equally essentially
it is the one and universal impulse of the specific, which
leads back into unity this its particularisation and there
preserves it."

P. 229 (1). Hegel, *Logik*, A., II, pp. 261-2, B., II, pp. 414-5 :
" According to this side the intro-Reflection of the Kind
is that by means of which it obtains actuality, because the
moment of negative unity and individuality is posited in it
—the propagation of living generations. The Idea (which,
as Life, is still in the form of immediacy) falls back into
actuality in so far, and this its Reflection is only repetition
and the infinite progress in which it does not emerge out of
the finitude of its immediacy. But this return into its
first Notion has also this higher side, that the Idea has
not only run through the mediation of its Processes within
immediacy, but also by this very fact has transcended
them and has thereby raised itself into a higher form of its
existence.

For the Process of the Kind (in which the single Individuals cancel in one another their indifferent and immediate
existence and die away in this negative unity) has further,
and for the other side of its product, the Realised Kind,
which has posited itself as identical with the Notion. In
the Process of Kind the separated individualities of individual
Life perish ; the negative identity in which the Kind returns
into itself is (*a*) the creation of individuality, but (*b*) is also its
transcendence : it is thus the self-coinciding Kind, or
universality, *which becomes for itself*, of the Idea. In the
process of generation the immediacy of living individuality
perishes : the death of this Life is the emergence of Spirit.
The Idea, which as Kind *in itself*, is *for itself*, since it has
transcended its particularity which constituted the living
generations, and has so given itself a reality, which is
itself simple universality : it is thus the Idea which is
related to itself as Idea, the universal which has universality
for its determinateness and existence, or, *the Idea of
Cognition*."

P. 229 (2). Hegel, *Enzyklopädie*, C., I, pp. 83-4, D.,
pp. 81-2 : " Touching this principle it has been justly observed

that in what we call Experience, as distinct from mere single perception of single facts, there are two elements. The one is the matter, infinite in its multiplicity, and as it stands a mere set of singulars : the other is the form, the characteristics of universality and necessity. Mere experience no doubt offers many, perhaps innumerable, cases of similar perceptions : but, after all, no multitude, however great, can be the same thing as universality. Similarly, mere experience affords perceptions of changes succeeding each other and of objects in juxtaposition ; but it presents no necessary connection. If perception, therefore, is to maintain its claim to be the sole basis of what men hold for truth, universality and necessity appear something illegitimate : they become an accident of our minds, a mere custom, the content of which might be otherwise constituted than it is."

P. 229 (4). Hegel, *Enzyklopädie*, C., I, pp. 26-7, D., p. 29: " In Nature nothing else would have to be discerned, except the Idea : but the Idea has here divested itself of its proper being. In Mind, again, the Idea has asserted a being of its own, and is on the way to become absolute."

P. 233. Hegel, *Logik*, A., I/2, pp. 205-6, B., II, p. 177 : " The Contingent therefore offers two sides. First, in so far as it immediately contains Possibility, or (which is the same thing) in so far as Possibility is in a transcended state in it, it is neither positedness nor is it mediated, but is immediate Actuality ; it has no Ground. This immediate Actuality belongs to the Possible too, and therefore it is determined equally as the Actual and as contingent, and thus is groundless as well.

But, secondly, the Contingent is the Actual as a *merely* Possible or as a positedness ; and similarly the Possible as formal Being-in-Self is merely positedness. Thus neither is in and for itself, but each has its veritable intro-Reflection in an Other : or, it has a Ground.

The Contingent, then, has no Ground because it is contingent ; and, equally, because it is contingent, it has a Ground."

P. 235. Heine, *Collected Works*, III, Romanzero, *Disput*, p. 186.

> Gilt nicht mehr der Tausves Jontof !
> Was soll gelten ?   Zeter !   Zeter !
> Rache, Herr, die Missetat,
> Strafe, Herr, den Übeltäter !

> (Now law and prophet get no heed !
> What shall prevail ?   O help !   O help !
> Revenge, O Lord, the evil deed,
> Punish, O Lord, the evil-doer !)

P. 240 (1). Hegel, *Logik*, A., II, pp. 35–6, B., II, pp. 234–5 : " Now this general Notion which is here to be considered contains the three moments of Universality, Particularity, and Individuality.   The distinction, and the determinations which result in the process of distinguishing, constitute the side which before was called positedness.   This is in the Notion identical with Being-in-and-for-Self, and therefore each of these moments is whole Notion as much as determinate Notion or as one determination of the Notion.

At first it is pure Notion or the determination of Universality.   But the pure or Universal Notion is also only a determinate or Particular Notion, which places itself alongside of the others.   The Notion is the totality, and thus in its universality or pure identical self-relation is essentially the fact of determining and distinguishing ; and therefore it contains in itself the standard by means of which this form of its self-identity, penetrating and comprehending all the moments, equally immediately determines itself to be *only* the Universal as against the distinctness of the moments.

Secondly the Notion is hereby as this Particular Notion, or as the determinate Notion which is posited as distinct from others.

Thirdly, Individuality is the Notion which reflects itself out of distinction into absolute negativity.   This at the same time is the moment in which it has passed out of its identity into its otherness, and becomes the Judgment."

P. 240 (2). Hofmann, *Ein Jahrhundert chemischer Forschung*, pp. 53–4 : " I have already had in mind the peculiar trend of mind which regrettably lamed scientific research in our fatherland during the first two decades of this century. Why should there be any need of observation ? These philosophers of nature indeed already knew everything, or if they did not know it, at least they did not lack for words to deceive themselves and others as to their absence of knowledge. We smile to-day at the exuberant phraseology in which they endeavoured to clothe the description of the simplest phenomenon, and at the fantastic metaphors of their would-be explanations, and it is difficult for us to understand how many men, in part highly talented, could find satisfaction during a long series of years in this barren trifling. And what at once estranges us is that it was precisely circles in Berlin in which these conceptions took deepest root."

P. 7–8 : " It is, of course, known far outside the narrow bounds of specialist circles that the first attempts to naturalise the winning of sugar from indigenous beets in our fatherland were carried out in the reign of Frederick Wilhelm III. But it was only in the most recent period that it has become common knowledge how greatly this naturalisation was facilitated and hastened by the personal action of the king ; it is only in the most recent period that the publication of official archives has convincingly proved at what an early stage the king granted his most energetic attention to these endeavours, and how during long and often troubled years he never tired of helping towards the success of these endeavours, the great import of which was indubitable to his clear view, by unremitting sympathy and perspicacious support.

P. 27 : Few at that time could still be inclined to regard diamonds as quartz that has come to consciousness of itself or ' platinum as the paradox of silver wishing to occupy the highest stage of metallicity, which belongs only to gold ' . . . [to which the following note] : ' To these few belonged as late as 1850 even *Karl Rosenkranz*, from whose ' *System der Wissenschaft. Ein philosophisches Encheiridion* ' the passage quoted in the text is taken. The passage says, p. 301, paragraph 475 : ' 3. The *noble* metals are the heaviest

and at the same time those which combine the most extreme
expansibility with the greatest contraction of their cohesion.
(*a*) Mercury repeats at this stage the peculiarity of the light
metals of fusing easily with other metals. Its dynamic
excitability is so great that it becomes liquid at heightened
temperature and solidifies only at a very low one. (*b*) *Silver*
is the higher reproduction of the base metals. Brittle,
dirty white *platinum* is, it is true, specifically somewhat
heavier than gold, but basically only a paradox of silver,
wishing to occupy already the highest stage of metallicity.
This belongs (*c*) only to *gold*, which as the heaviest metal
is at the same time the most expansible, and with the
exception of chlorine is unattacked by all chemical powers.
In it metallicity reaches complete saturation and hence
also it encounters us laughing with the warm shine of a pure
lovable yellow.' "

P. 248 (1). Hegel, *Logik*, A., I/1, p. 268, B., I, p. 245 :
" And what causes thought to fail and produces the fall and
the dizziness is nothing else than the weariness of repetition,
where the limit is made to vanish, return, and vanish again
perpetually, and the hither to arise and perish in the beyond,
and the beyond in the hither, one after the other, and the
one in the other ; which only causes a feeling of the impotence
of this infinite and this *ought*, which would master the finite
and cannot."

P. 248 (2). Hegel, *Enzyklopädie*, C., p. 57, D., p. 56: "When
the record adds that God drove men out of the Garden of
Eden to prevent their eating of the tree of life, it only means
that on his natural side certainly man is finite and mortal,
but in knowledge infinite."

P. 249. Hegel, *Logik*, A., I/1, p. 236-7, B., I, p. 220 :
" Arithmetic contemplates Number and its figures ; or,
rather, it operates with them and does not contemplate.
For Number is indifferent determinateness and inert ; it
must be actuated and brought into relation from without.
The different methods of relation are the species of calculation. In arithmetic they are enumerated in series, and
it is evident that they are mutually dependent. But arith-

metic does not give prominence to the thread which guides their progress. However, the systematic arrangement which the exposition of these elements in the text-books justly claims, results easily from the conceptual determination of Number itself."

P. 256. Wiedemann, II, p. 635–6 : " The introduction of a function which coincides with Weber's law at a finite distance but deviates from it at *molecular* distances cannot, therefore, solve the contradictions under consideration."

P. 257. Hegel, *Enzyklopädie*, C., I, pp. 205–7, D., p. 193 : " Intensive magnitude or Degree is in its notion distinct from Extensive magnitude or the Quantum. It is therefore inadmissible to refuse, as many do, to recognise this distinction, and without scruple to identify the two forms of magnitude. They are so identified in physics, when difference of specific gravity is explained by saying, that a body, with a specific gravity twice that of another, contains within the same space twice as many material parts (or atoms) as the other. So with heat and light, if the various degrees of temperature and brilliancy were to be explained by the greater or less number of particles (or molecules) of heat and light. No doubt the physicists, who employ such a mode of explanation, usually exclude themselves when they are remonstrated with on its untenableness, by saying that the expression is without prejudice to the confessedly unknowable essence of such phenomena, and employed merely for greater convenience. This greater convenience is meant to point to the easier application of the calculus : but it is hard to see why Intensive magnitudes, having, as they do, a definite numerical expression of their own, should not be as convenient for calculation as Extensive magnitudes. If convenience be all that is desired, surely it would be more convenient to banish calculation and thought altogether. A further point against the apology offered by the physicists is, that to engage in explanations of this kind is to overstep the sphere of perception and experience, and to resort to the realm of metaphysics and of what at other times would be called idle or even pernicious speculation. It is certainly a fact of experience that, if one of two purses filled with shillings is twice as heavy as the other, the reason must be, that the one contains, say two hundred, and the other only one hundred shillings. These pieces of money we can see and feel with our senses : atoms, molecules, and the like, are on the contrary beyond the range of sensuous perception, and thought alone can decide whether they are admissible, and have a meaning."

P. 258. Mädler, p. 44 : " The most important passage that has come down to us in this connection is that of *Aristarchus* of Samos. ' The earth,' he says, ' revolves on its axis and at the same time in an inclined circle round the sun. This circle, however, in relation to the distance of the fixed stars has only the relation of the centre to the periphery, and consequently we cannot perceive the motion of the earth by the fixed stars.' Certainly the truest and most correct thing that could be said at all in an epoch when it had not at all been made clear what it really was that had to be explained."

P. 259 (1). Wolf, *Geschichte der Astronomie*, p. 313 : " In fact, at that time he had not only seen mountains in the moon, but had even already attempted to determine the height of some of them—he had distinguished forty stars in the Pleiades, he had discovered some other similar aggre-gations of stars in Orion, Cancer, etc., and had recognised that the shining of the Milky Way, attributed even by Aristotle to meteors, in accordance with the guess already put forward by *Democritus*, was the united light of in-numerable small stars. . . ."

P. 259 (2). Guthrie, *Magnetism and Electricity*, p. 263–4 : " Let us for brevity call a circular current $+$ when it moves round from our aspect in the direction of the hands of a watch, and $-$ when in the opposite direction. Then the current around the S. pole when it faces us is $+$, that around the N. pole is $-$. Then the attraction between N. and S., fig. 238, is due to the neighbouring and prevailing parts of the currents moving in the same direction. With like poles, N. and N'., and S. and S'., the neighbouring currents are moving in opposite directions, and there is, accordingly, repulsion. This attraction or repulsion extends along the whole length of two magnets put side by side, fig. 238, where the letters u and d denote whether the current is emerging from, or entering the plane of the paper. Notice here the entire similarity between this figure and fig. 213, Art. 267, where spiral currents alone acted on one another. Further, if (fig. 239) the unlike ends of two magnets are presented to one another, they attract, because their Ampèreian currents are parallel and in the same direction. Like poles repel, because their currents are in opposite directions."

P. 264. *Grove*, p. 211 : " The voltaic battery affords us the best means of ascertaining the dynamic equivalents of different forces, and it is probable by its aid that the best theoretical and practical results will be ultimately attained."

P. 264. Thomson, Thomas, p. 358 : " Coulomb demonstrated that it is a consequence of the law, that the particles of electricity repel each other inversely as the square of their distance, that the electricity when accumulated in a conducting body is distributed totally on the surface of the body, and that none of it exists in the interior of the body. He showed likewise the truth of this law experimentally."

P. 264. Thomson, Thomas, p. 366 : " This subject attracted the attention of M. Poisson, who applied to it all the resources of the most refined calculus, and determined the thickness of the coating of electricity upon bodies of different forms from the hypothesis that positive and negative electricity are two fluids, the particles of each of which repel each other with forces varying inversely as the square of the distance ; while the vitreous electricity attracts the resinous with forces varying according to the same law. He showed that the exterior surface of the electrical coating coincides with that of the body, and that as the coating is very thin, the interior surface is but little distant from it. In a sphere both the exterior and interior surfaces are spherical, and the centre of these surfaces is the same with that of the centre of the body."

P. 265.   Thomson, Thomas, p. 360 : " Electricity, then, is all deposited on the surface of bodies, and the only obstacle to its leaving that surface and being instantly dissipated, is the pressure of the atmosphere."

P. 265.   Thomson, Thomas, p. 378–9 : " Faraday has drawn the following conclusions from his experiments :—

1. All bodies conduct electricity in the same way from metals to lac and gases, but in very different degrees.

2. Conducting power is in some bodies powerfully increased by heat, and in others diminished, yet without our perceiving any accompanying essential electrical difference either in the bodies, or in the changes occasioned by the electricity conducted.

3. A numerous class of bodies, insulating electricity of low intensity when solid, conduct it very freely when fluid, and are then decomposed by it.

4. But there are many fluid bodies which do not sensibly conduct electricity of this low intensity ; there are some which conduct it without being decomposed ; nor is fluidity essential to decomposition.

5. There is but one body yet discovered, the periodide of mercury, which, insulating a voltaic current when solid and conducting it when fluid, is not decomposed in the latter case.

6. There is no strict electrical distinction of conduction, which can yet be drawn between bodies supposed to be elementary, and those known to be compound."

P. 265.   Thomson, Thomas, p. 400 : " ' The spark is a discharge or lowering of the polarised inductive state of many dielectric particles, by a particular action of a few of the particles occupying a very small and limited space.'   Faraday conceives that ' the few particles where the discharge occurs are not merely pushed apart, but assume a peculiar state, a highly exalted condition for the time ; that is to say, have thrown upon them all the surrounding forces in succession, and rising up to proportionate intensity of condition perhaps equal to that of chemically combining atoms, discharge the powers, possibly in the same manner as they do theirs, by some operation at present unknown to us ; and so the end of the whole. The ultimate effect is exactly as if a metallic particle had been put into the place of the discharging particle ; and it does not seem impossible that the principles of action, in both cases, may hereafter prove to be the same.' I have given this explanation of Faraday in his own words, because I do not clearly understand it."

P. 265. Thomson, Thomas, p. 405 : " When a given quantity of electricity occasions a spark by passing from one body to another, its brilliancy is always the greater the smaller the size of the body from which it is drawn ; hence it happens that more brilliant sparks may be drawn from a small brass knob, fixed to the prime conductor of an electrical machine, than from the prime conductor itself. A short spark is always white ; but a very long spark is usually reddish, or rather purplish. When we draw a spark from the prime conductor of an electric machine, by means of a metallic knob, the spark is white ; but when we draw it by the hand it is purple. If we draw it by means of a wet plant, or water, or ice, the colour is red. The same spark will vary in colour according to its length. When short it is always white, when very long it is purple or violet. A spark which in the open air does not exceed a quarter of an inch in length, will appear to fill the whole of an exhausted receiver, four inches wide, and eight inches long. In the former case it is white, in the latter the light is very feeble, and the colour violet."

P. 266 (1). Thomson, Thomas, p. 409–10 : " It was an opinion maintained about thirty years ago, by many eminent experimenters in Germany, that the electric light is of the same nature with fire, and that it is formed by the union of the two electricities. This opinion appears to have been first stated by Winterl ; and, unless I misunderstand Ritter, he seems to have entertained the same sentiments. But this opinion, though it appears at first sight plausible— and though it would be very convenient to be able to account so well for the analogy which obviously exists between fire and electricity—will not bear a rigid examination. Every person who has seen an electric spark, must be aware that the passage is so instantaneous that it is impossible to say from which point it proceeds, or to which it goes. If the spark be long, that is to say if the distance between the two knobs between which it passes be considerable, the presence of the two kinds of electricity may be at once observed. Suppose one of the knobs attached to the prime conductor of an electrical machine, and the other attached to a conduction body connected with the earth—the portion of the spark nearest the prime conductor of the machine exhibits all the characters which distinguish positive electricity—while the portion of the spark nearest the other knob, exhibits the characters of negative electricities. When two charged bodies are placed within the striking distance, no spark will pass unless the one body be charged with positive, and the other with negative electricity. The two electricities are attracted towards each other, advance at the same instant from each of the charged bodies, and uniting together somewhere between the two knobs, all symptoms of electricity are at an end. When a spark is short, the whole distance between the two knobs through which it passes, is equally illuminated; but when the spark is long, those portions of it which are next the knobs, are much brighter than towards the centre of the spark. Near the knobs the colour is white, but towards the centre of the spark purplish. Indeed, if the spark be very long, the middle part of it is not illuminated at all, or only very slightly. Now, this imperfectly illuminated part is obviously the spot where the two electricities unite, and it is in consequence of this union that the light is so imperfect."

P. 266 (2). Thomson, Thomas, pp. 415–16 : " With respect to the kind of electricity, M. Dessaignes found that when the mercury in the barometer is rising, and the temperature of the atmosphere becoming colder, glass, amber, and sealing-wax, cotton, silk, and linen, when plunged into mercury, are always negative ; but they are positive when the barometer is falling, and the atmosphere becoming warmer. Sulphur was always positive. During summer he always found these bodies positive in impure, and negative in pure, mercury.

Cold, as well as heat, destroys the electricity in these experiments."

P. 266 (3). Thomson, Thomas, p. 419: "In order to produce thermo-electric effects, it is not necessary to apply heat. Anything which alters the temperature in one part of the chain, from that of the rest, occasions a deviation in the declination of the magnet : for example, if we produce cold in any part of the antimony bar, by applying ether to it, and allowing it to evaporate ; or if we cool it by the application of ice. The greatest effect of all is produced on the magnet, when one part of the bar is heated, and the other cooled. It is evident from this, that the evolution of electricity depends upon the difference in the temperature of different parts of the metallic chain."

P. 266 (4). Thomson, Thomas, pp. 437–8: "Now, as bodies are attracted by those in a different state of excitement from themselves, it follows that oxygen, chlorine, bromine, and iodine, and acids, would not be attracted to the positive pole, unless they themselves were in a negative state ; nor would hydrogen and bases be attracted to the negative pole unless they were in a positive state. From this it has been concluded that bodies which have an attraction for each other are in opposite states of electricity, and that it is to these opposite states that their attraction for each other, and their union with each other is owing. The current of electricity destroys their union by bringing them into the same electrical state. In consequence of this view, which is at least exceedingly ingenious and plausible, bodies have been divided into two sets, those which are negative, and those which are positive."

P. 267. Wiedemann, II, p. 418 : "These phenomena have an interest that is more chemical than physical ; consequently we have only briefly mentioned them."

P. 301. Wallace, p. 229 : "In the following passage from Iamblichus on Divination, quoted in Maurice's *Moral and Metaphysical Philosophy*, we find mention in a short space of a number of the most startling phenomena of modern Spiritualism :
' Often at the moment of inspiration, or when the afflatus has subsided, a fiery appearance is seen—the entering or

departing power. Those who are skilled in this wisdom can tell by the character of this glory the rank of the divinity who has seized for the time the reins of the mystic's soul, and guides it as he will. Sometimes the body of the man is violently agitated, sometimes it is rigid and motionless. In some instances sweet music is heard, in others discordant and fearful sounds. The person of the subject has been known to dilate and tower to a superhuman height, in other cases it has been lifted into the air. Frequently not merely the ordinary exercise of reason, but sensation and animal life would appear to have been suspended; and the subject of the afflatus has not felt the application of fire, has been pierced with spits, cut with knives, and not been sensible to pain.' "

P. 302. Wallace, pp. 187–8 : " The accounts of spirit-photography in several parts of the United States caused many spiritualists in this country to make experiments, but for a long time without success. Mr. and Mrs. Guppy, who are both amateur photographers, tried at their own house, and failed. In March, 1872, they went one day to Mr. Hudson's, a photographer living near them (not a spiritualist) to get some *cartes de visite* of Mrs. Guppy. After the sitting the idea suddenly struck Mr. Guppy that he would try for a spirit-photograph. He sat down, told Mrs. G. to go behind the background, and had a picture taken. There came out behind him a large, indefinite, oval, white patch, somewhat resembling the outline of a draped figure. Mrs. Guppy, behind the background, was dressed in black. This is the first spirit-photograph taken in England, and it is perhaps more satisfactory on account of the suddenness of the impulse under which it was taken, and the great white patch which no impostor would have attempted to produce, and which taken by itself, utterly spoils the picture. A few days afterwards, Mr. and Mrs. Guppy and their little boy went without any notice. Mrs. G. sat on the ground holding the boy on a stool. Mr. Guppy stood behind looking on. The picture thus produced is most remarkable. A tall female figure, finely draped in white, gauzy robes, stands directly behind and above the sitters, looking down on them and holding its open hands over their heads, as if giving a benediction. The face is somewhat Eastern, and, with the hands, is beautifully defined. The white robes pass behind the sitters' dark figures without in the least showing through. A second picture was then taken as soon as a plate could be prepared, and it was fortunate it was so, for it resulted in a most remarkable test. Mrs. G. again knelt with the boy ; but this time she did not stoop so much, and her head was higher. The same white figure comes out equally well defined, but it has changed its position in a manner exactly corresponding to the slight change of Mrs. G.'s position. The hands were before on a level ; now one is raised considerably higher than the other, so as to keep it about the same distance from Mrs. G.'s head as it was before. The folds of the drapery all correspondingly differ, and the head is slightly turned. Here, then, one of two things are absolutely certain. Either there was a living, intelligent, but invisible being present, or Mr. and Mrs. Guppy, the photographer, and some fourth person, planned a wicked imposture, and have maintained it ever since. Knowing Mr. and Mrs. Guppy so well as I do, I feel an absolute conviction that they are as incapable of an imposture of this kind as any earnest inquirer after truth in the department of natural science."

P. 306.   Davies, *Mystic London*, p. 319 : " As a final *bonne bouche* the spirit made its exit from the side of the folding door covered by the curtain, and immediately Miss C. rose up with dishevelled locks in a way that must have been satisfactory to anybody who knew nothing of the back door and the brawny servant, or who had never seen the late Mr. Charles Kean act in the ' Corsican Brothers ' or the ' Courier of Lyons.'

I am free to confess, the final deathblow to my belief that there might be ' something in ' the Face Manifestations was given by the effusive Professor who has ' gone in ' for the Double with a pertinacity altogether opposed to the calm judicial examination of his brother learned in the law, and with prejudice scarcely becoming a F.R.S."

P. 310.   Huxley, Letter to the Committee of the London Dialectical Society, *Daily News*, 17, X, 1871.

BIBLIOGRAPHY

ALEMBERT, d' (Jean le Rond). *Traité de dynamique,* Paris, David l'aîné, 1743.

ALLMAN, G. J. Recent progress in our knowledge of the ciliate infusoria. *Nature,* Vol. XII, pp. 136–7, 155–7, 175–7, 1875.

BOSSUT, Charles. *Traités de calcul différentiel et de calcul intégral.* 2 vols. Vol. I, Paris, L'imprimerie de la République, 1795.

BÜCHNER, Louis. A. *Kraft und Stoff.* 7th edition. Leipzig, Theodor Thomas, 1862.

B. —— English translation : *Force and Matter.* Translated by J. F. Collingwood, 2nd edition, completed from the 10th German edition, London, 1870.

CARNOT, S. *Réflexions sur la puissance motrice du feu et sur les machines propres à développer cette puissance.* Paris, Bachelier, 1824.

CLAUSIUS, R. A. *Über den zweiten Hauptsatz der mechanischen Wärmetheorie.* Braunschweig, Fr. Vieweg und Sohn, 1867.

B. *Die mechanische Wärmetheorie.* 2nd edition, Vol. I, Braunschweig, Fr. Vieweg und Sohn, 1876.

CROLL, James. *Climate and Time in their Geological Relations.* London, Daldy, Ibister and Co., 1875. Reviewed by J. F. B. in *Nature,* Vol. XII, pp. 121–3, 141–4.

DARWIN, Charles, *The Origin of Species.* 6th edition. London, John Murray, 1873.

DAVIES, Charles Maurice, *Mystic London.* London, Tinsley Brothers, 1875.

DRAPER, John William. *History of the Intellectual Development of Europe.* 2 vols. London, Bell and Daldy, 1864.

FARADAY, Michael. *Experimental Researches in Electricity.* 3 vols. Vol. I. London, Bernard Quaritch, 1839.

FICK, Adolf. *Die Naturkräfte in ihrer Wechselbeziehung.* Popular lectures. Würzburg, Stahel, 1869.

GALIANI, Ferdinando. *Della moneta.* In : *Scrittori classici italiani di economia politica.* Vol. IV. Milano, Nella Stamperia e Fonderia di G. G. Destefanis, 1803.

GERLAND, E. *Leibnizen's und Huygens' Briefwechsel mit Papin.* Berlin, 1881.

GROVE, W. R. *The Correlation of Physical Forces.* 3rd edition. London, Longman, Brown, Green and Longmans, 1855.

GUTHRIE, Frederick. *Magnetism and Electricity.* In : Collin's Advanced Science Series. London and Glasgow, William Collins, Sons and Co., 1876.

HÆCKEL, Ernst. A. *Anthropogenie oder Entwickelungsgeschichte des Menschen.* Leipzig, Wilhelm Engelmann, 1874.

B. *Natürliche Schöpfungsgeschichte.* 5th edition. Berlin, Georg. Reimer, 1874.

C. —— English Translation : *The History of Creation,* revised by Professor E. Ray Lankester. 2 vols. H. S. King & Co., London, 1876.

D. *Die Perigenesis der Plastidule oder die Wellenzeugung der Lebensteilehen.* Berlin, Georg. Reimer, 1876.

HEGEL, Georg. Wilhelm Friedrich. Collected Works. Berlin, Duncker und Humblot, 1832–45.

A. *Wissenschaft der Logik.* Vols. III–V. 1833–4.

B. —— English translation : *Science of Logic.* 2 Vols. London, Allen and Unwin, 1929.

C. *Enzyclopädie der philosophischen Wissenschaften im Grundrisse.* Part I. *Logik.* Vol. VI. 1842.

D. —— English translation : *The Logic of Hegel.* Translated by W. Wallace. 2nd edition. Oxford, Clarendon Press, 1892.

E. *Enzyclopädie der philosophischen Wissenschaften im Grundrisse.* Part II. *Vorlesungen uber die Naturphilosophie.* Vol. XII. 1842.

F. *Vorlesungen uber die Geschichte der Philosophie.* Vols. XIII–XV. 1833–6.

HELMHOLTZ, H. A. *Populäre wissenschaftliche Vorträge.* Sheets I–III. Braunschweig, Friedrich Vieweg und Sohn, 1865, 1871, 1876.

B. *Über die Erhaltung der Kraft.* Berlin, G. Reimer, 1847.

HOFMANN, Aug. Wilhelm. *Ein Jahrhundert chemischer Forschungen unter dem Schirme der Hohenzollern.* In : *Chemische Erinnerungen aus der Berliner Vergangenheit.* Berlin, August Hirschwald, 1882.

KANT, Immanuel. Collected Works, Berlin, Georg Reimer, 1910–11.

A. Vol. I. *Gedanken von der Wahren Schätzung der lebendigen Kräfte,* 1747.

B. Vol. I. *Untersuchung der Frage ob die Erde in ihrer Umdrehung um die Achse einige Veränderung seit den ersten Zeiten erlitten habe.* 1754.

C. Vol. I. *Allgemeine Naturgeschichte und Theorie des Himmels.* 1755.

D. Vol. III. *Kritik der reinen Vernunft.* 1787.

KEKULÉ, August. *Die wissenschaftliche Ziele und Leistungen der Chemie.* Bonn, Max Cohen und Sohn, 1878.

KIRCHHOFF, Gustav. *Vorlesungen über mathematische Physik. Mechanik.* Leipzig, B. G. Teubner, 1876.

KOHLRAUSCH, F. *Das electrische Leitungsvermögen der wässerigen Lösungen von den Hydraten und Salzen der leichten Metalle. sowie von Kupfervitriol, Zinkvitriol und Silbersalpeter.* In : *Wiedemann's Annalen der Physik und Chemie.* New Series, Vol. VI, No. 1. Leipzig, Barth, 1829.

KOPP, Hermann. *Die Entwicklung der Chemie in der neueren Zeit. In Geschichte der Wissenschaften in Deutschland.* Vol. X, Part 1. München, R., Oldenbourg, 1871.

LAPLACE, P. S. *Traité de méchanique celeste.* 5 vols. Paris, I. B. M. Duprat, 1796.

LAVROV, P. L. *Opyt istorii mysli.* (Attempt at a History of Thought.) Vol. I. St. Petersburg, published by the journal *Znanye,* 1875.

LIEBIG, Justus von. *Chemische Briefe.* Leipzig und Heidelberg, C. F. Winter, 1865.

LUBBOCK, John. *Ants, Bees and Wasps.* London, Kegan Paul, Trench & Co., 1882. Reviewed by G. J. Romanes in *Nature,* Vol. XXVI, pp. 121–3.

MÄDLER, J. H. *Der Wunderbau des Weltalls, oder populäre Astronomie.* 5th edition. Berlin, Carl Heymann, 1861.

MAXWELL, J. Clerk. *Theory of Heat.* London. Longmans, Green & Co. 1871.

MAYER, J. R. A. *Bemerkungen über die Kräfte der unbelebten Natur.* In : *Annalen der Chemie und Pharmacie,* von Fr. Wöhler und J. Liebig, Vol. XLII. Heidelberg, C. F. Winter, 1842.

B. *Die Mechanik der Wärme.* In : *Collected Writings.* Stuttgart. J. G. Cotta, 1867.

MEYER, Lothar. *Die Natur der chemischen Elemente als Funktion ihrer Atomgewichte.* In : Liebig's *Annalen der Chemie.* Supplement, Vol. VII. Leipzig, C. F. Winter, 1870.

NÄGELI, C. von. *Über die Schranken der naturwissenschaftlichen Erkenntnis.* In the official report of the fiftieth meeting of German scientists in Munich. München, F. Straub, 1877.

NAUMANN, Alexander. *Handbuch der allgemeinen und physikalischen Chemie.* Heidelberg, Carl Winter, 1877.

OWEN, Richard. *On the Nature of Limbs.* London, 1849.

ROSCOE, H. E.—SCHORLEMMER, C. *Ausführliches Lehrbuch der Chemie.* 9 vols. Braunschweig, Fr. Vieweg und Sohn. Vol. I. Non-Metals, 1877. Vol. II. Metals and Spectral Analysis, 1879.

SECCHI, P. A. *Die Sonne.* German edition. 2 vols. Braunschweig, G. Westermann, 1872.

SUTER, H. *Geschichte der mathematischen Wissenschaften.* Part 2· Zürich, Orell Füssli & Co. 1815.

TAIT, P. G. *Force.* Nature, Vol. XIV, pp. 459–63. 1876.

THOMSON, Thomas. *An Outline of the Sciences of Heat and Electricity.* 2nd edition. London, Baillière, 1840.

THOMSON, William. *The Size of Atoms.* Nature, Vol. XXVIII, pp. 203–5, 250–4, 274–8, 1883.

THOMSON, W., and TAIT, P. G. A. *Treatise on Natural Philosophy.* Vol. I, Oxford, Clarendon Press, 1867.

B. *Handbuch der theoretischen Physik.* German translation by H. Helmholtz and G. Wertheim, Vol. I, part 2. Braunschweig, F. Vieweg und Sohn, 1874. Preface by H. Helmholtz.

TYNDALL, John. A. *Inaugural Address to the 44th meeting of the British Association in Belfast, Nature.* Vol. X, pp. 309–19.

B. *On the optical deportment of the atmosphere in reference to the phenomena of putrefaction and infection.* In: Nature, Vol. XIII, pp. 252–4, 268–70, 1876.

VIRCHOW, Rudolf. *Die Zellularpathologie.* In *Vorlesungen über Pathologie.* Vol. I. 4th edition. Berlin, Hirschwald, 1871.

WAGNER, Moriz. *Naturwissenschaftliche Streitfragen.* I. J. von Liebig's Views on the Origin of Life and the Theory of Descent. In Supplement to *Allgemeine Zeitung,* Nos. 279, 280, 281. Augsburg, J. G. Cotta, 1874.

WALLACE, Alfred Russell. *On Miracles and Modern Spiritualism,* 3 essays. London, James Burns, 1875.

WHEWELL, William. *History of the Inductive Sciences.* 3rd edition. 3 vols. London, J. W. Parker & Son, 1857.

WIEDEMANN, Gustav. *Die Lehre vom Galvanismus and Elektro-magnetismus.* 2nd edition. 2 vols. Braunschweig, F. Vieweg und Sohn, 1872–4.

WOLF, Rudolf. *Geschichte der Astronomie.* In : *Geschichte der Wissenschaften in Deutschland.* Vol. XVI. München, Oldenburg, 1877.

WUNDT, Wilhelm. *Lehrbuch der Physiologie des Menschen.* 2nd edition. Erlangen, Enke, 1868.

INDEX